T0142796

Springer Theses

Recognizing Outstanding Ph.D. Research

Aims and Scope

The series "Springer Theses" brings together a selection of the very best Ph.D. theses from around the world and across the physical sciences. Nominated and endorsed by two recognized specialists, each published volume has been selected for its scientific excellence and the high impact of its contents for the pertinent field of research. For greater accessibility to non-specialists, the published versions include an extended introduction, as well as a foreword by the student's supervisor explaining the special relevance of the work for the field. As a whole, the series will provide a valuable resource both for newcomers to the research fields described, and for other scientists seeking detailed background information on special questions. Finally, it provides an accredited documentation of the valuable contributions made by today's younger generation of scientists.

Theses are accepted into the series by invited nomination only and must fulfill all of the following criteria

- They must be written in good English.
- The topic should fall within the confines of Chemistry, Physics, Earth Sciences, Engineering and related interdisciplinary fields such as Materials, Nanoscience, Chemical Engineering, Complex Systems and Biophysics.
- The work reported in the thesis must represent a significant scientific advance.
- If the thesis includes previously published material, permission to reproduce this must be gained from the respective copyright holder.
- They must have been examined and passed during the 12 months prior to nomination.
- Each thesis should include a foreword by the supervisor outlining the significance of its content.
- The theses should have a clearly defined structure including an introduction accessible to scientists not expert in that particular field.

More information about this series at http://www.springer.com/series/8790

Vera M. Schäfer

Fast Gates and Mixed-Species Entanglement with Trapped Ions

Doctoral Thesis accepted by
University of Oxford, Oxford, UK

Author
Dr. Vera M. Schäfer
Clarendon Laboratory
Department of Physics
University of Oxford
Oxford, UK

Supervisor
Prof. David Lucas
Clarendon Laboratory
Department of Physics
University of Oxford
Oxford, UK

ISSN 2190-5053 ISSN 2190-5061 (electronic)
Springer Theses
ISBN 978-3-030-40287-7 ISBN 978-3-030-40285-3 (eBook)
https://doi.org/10.1007/978-3-030-40285-3

This Springer imprint is published by the registered company Springer Nature Switzerland AG
The registered company address is: Gewerbestrasse 11, 6330 Cham, Switzerland

Supervisor's Foreword

Vera Schäfer's thesis reports exciting developments in the field of experimental quantum computing, using qubits (quantum bits) stored in single atomic ions.

Various different technologies are being explored for future quantum computers, which promise to be able to tackle certain computational problems vastly more efficiently than conventional computers. Amongst these technology platforms, qubits based on trapped ions, confined in high vacuum by electric fields and manipulated by pulses of laser light, were one of the first systems to be proposed for quantum information processing; they were also the first technology in which quantum logic gates were demonstrated, initially by the group of David Wineland at NIST Boulder in the USA, in the late 1990s. The "two-qubit" logic gate may be seen as the fundamental operation of quantum computing, because it generates an entangled quantum state between two qubits which are initially non-entangled, and entanglement is one of the principal features which sets quantum computers apart from their conventional counterparts.

Trapped atomic ions make exceptionally good qubits in several ways: they are extremely good quantum memories, in that a qubit's quantum state can be preserved against decoherence for timescales of many minutes; the elementary logic operations can be made very precise, which is essential for quantum computing; and atoms also couple naturally to visible light photons, giving the potential for quantum networking via optical fibres. However, one "Achilles' Heel" that they have suffered from is that the fundamental two-qubit logic operations are rather slow. Although the precision of these logic gates has been improved by more than two orders of magnitude since the initial demonstrations twenty years ago, the speed had remained roughly the same.

The timescale that sets the speed of such gates is the natural oscillation period of ions in the trap: like a swinging pendulum, or a ball bearing rolling in a bowl, trapped ions can oscillate freely in the harmonic potential set up the electric fields of the trap. The stronger the electric field, and the lighter the ion, the faster will be the frequency of oscillation. For typical atomic ions and field strengths conveniently attainable in the laboratory, this frequency is in the few-megahertz range. All previous logic gates had operated in an adiabatic regime in which the gate operation needs to be much slower than this natural timescale, leading to typical operation times for gates of around 50 μs, translating to a rather slow "clock speed" of only 20 kHz or so. This is certainly not a fundamental limitation: the ions are only a few microns apart and the Coulomb interaction on which the gate operation relies propagates at the speed of light. Starting around 2003, there have been various theoretical proposals for accelerating the gate speed even beyond the natural timescale of the ion motion. However, none of these had been successfully implemented prior to the work reported in Vera Schäfer's thesis.

Conventional logic gates in ion traps are usually implemented with a single laser pulse, whose frequency and amplitude control the gate speed. Increasing the laser intensity makes the gate faster, but if this is done naïvely then several sources of error creep in. The main new ingredient in the logic gate method that Vera Schäfer demonstrated was to replace the single laser pulse by a pulse whose amplitude is shaped in time. The pulse shape is still fairly simple, consisting of 5 or so "steps" of specific amplitude and duration, but these new degrees of freedom allow one to compensate for many of the errors that arise from trying to drive the gate faster. This idea was theoretically proposed in 2014 by Andrew Steane and colleagues, and can allow gates to be performed at timescales comparable to the natural timescale of the ion motion, i.e. at megahertz speeds. One of the challenges in the laboratory was achieving precise control of the pulse shape, for example, the necessary sub-nanosecond timing precision. Initial results showed that the method worked, but the precision of the gate operation was not as high as predicted by the theory; after painstakingly characterizing a number of possible technical sources of error, Vera and her colleagues were confident that the apparatus was performing as intended. The theory was revisited and we discovered that one of the error sources had been underestimated; with the theory extended, and the experimental parameters adjusted accordingly, experiment and theory matched within the estimated uncertainties.

Importantly, with the revised theory available to guide the choice of the optimum experimental parameters, it was possible to demonstrate a logic gate with comparable precision (99.8%) to state-of-the-art conventional gates, but more than an order of magnitude faster (at only 1.6 μs). The fastest gate demonstrated was 480 ns, comparable with the speed of some solid state qubit systems, and at a timescale slightly shorter than the period of oscillation of an ion in the trap: this led to a description of the work as "breaking the sound barrier" of the system, in a companion piece published alongside the Letter reporting the results in *Nature* magazine (2018).

Although the research on fast gates is the main focus of the thesis, Vera Schäfer also did ground-breaking work on "mixed-species" logic gates, *viz.*, operations both between different isotopes of calcium ($^{43}Ca^+$ and $^{40}Ca^+$), and between different atomic elements (calcium and strontium). In the mixed-isotope case, she made one of the first tests of the quantum mechanical Bell inequality using different types of atom. Mixed-species gates are likely to form an important tool in scaling up trapped-ion quantum processors to larger numbers of qubits, as they allow the interfacing of different types of ions which can be used for distinct purposes.

The logic gates demonstrated might be fast; of course, the careful research needed to produce them was not! The work was a beautiful demonstration of the interplay of theory and experiment, made possible by the well-understood physics of atom/laser interactions in the low-noise environment of an ion trap. Vera Schäfer's thesis describes it with a clarity and rigour that it richly deserves.

Oxford, UK   Prof. David Lucas
December 2019

Abstract

Building a quantum computer is one of the outstanding contemporary goals in physics. Trapped-ion qubits are among the most promising contenders, achieving the highest precision gates out of all platforms. However, entangling gates have previously only been performed in a regime where their speed is limited by the motion of the ions in the trap. The work in this thesis demonstrates entangling operations beyond this speed limit. In addition, entanglement of different species of ion has been performed. This is an important step towards building a scalable quantum computer, by being able to interface and connect trapped ion qubits via photonic links.

We demonstrate mixed-species entangling gates using a method that requires only a single laser for both species. In an experiment demonstrating the technique with two isotopes of calcium, $^{40}Ca^+$ and $^{43}Ca^+$, we achieved a Bell state fidelity of $\mathscr{F} = 99.8(6)\%$. We also perform a Bell test with the two different isotopes, achieving a violation of the CHSH inequality by 15 standard deviations with CHSH parameter $S = 2.228(15)$, without making a fair-sampling assumption. We then use the same gate mechanism to perform an entangling gate on two different atomic species, $^{43}Ca^+$ and $^{88}Sr^+$. In a proof-of-principle experiment we achieved a Bell state fidelity of $\mathscr{F} = 83(3)\%$.

For achieving faster two-qubit gates we shape the amplitude of the laser pulses driving the gates. The pulse shapes are specifically designed to yield low error rates and be insensitive to the optical phase of the driving field [1]. We perform an entangling gate with fidelity $\mathscr{F} = 99.78(3)\%$ in $t_g = 1.6\,\mu s$, a factor of $20-60$ times faster than the previously highest-fidelity trapped ion gates with only a factor ≈ 2 increase in gate error. We also demonstrate entanglement generation within $480\,ns$, less than a single period of the centre-of-mass motion of the ions.

Acknowledgements

I first wish to thank my supervisor Prof. David Lucas for his great guidance, his enthusiasm and attention to detail, always taking time to give feedback or answer questions and careful proof-reading of this thesis.

I am very grateful to Dr. Chris Ballance, who has been superbly supervising me since my early days as a master's student, for his infectious enthusiasm and optimism, never growing tired of sharing his vast knowledge and understanding of physics and the experiment, and possibly even for his motivating sense of timing. Thank you for letting me do and learn things for myself and yet always having some helpful ideas when I get stuck, and also for proof-reading this thesis.

I wish to thank Prof. Andrew Steane for developing the theory for the fast gates and sharing his deep insight into physics, and Prof. Derek Stacey for giving me great advice on writing my first paper, teaching and atomic physics. Thank you to Keshav Thirumalai for building the trap, vacuum system and Strontium laser system and always being a sociable member of our group, and to Laurent Stephenson for helping to set up the new laser system and writing a much-improved new wavemeter interface. Thanks also go to Dr. Tom Harty for sharing his knowledge on electronics and programming, to Sarah Woodrow for designing the trap, to the newer eager members of the "old lab", David Nadlinger, Dr. Tim Ballance and Beth Nichols, and to all other group members past and present: Dr. Martin Sepiol, Dr. Norbert Linke, Dr. Hugo Janacek, Dr. Diana Aude Craik, Dr. Joe Goodwin, Jamie Tarlton, Jochen Wolf, Amy Hughes and Clemens Löschnauer. Thank you also to master's students Benjamin Merkel and Rustin Nourshargh for building and testing the B-field stabilisation and for making the rack-mounted lasers such a success, as well as to Matt Newport and Simon Moulder for machining many useful things in the workshop.

Danke an meine Eltern, Monika und Rudolf, und meine Geschwister Timo und Mina, für eure liebevolle Unterstützung. Danke Rudolf, dass du mir von klein an gezeigt hast wieviel Spass man mit Technik haben kann, und Monika, dass du die Physikbücher, die auf einmal aus der Bücherei aufgetaucht sind, entdeckt hast und mich ermutigt hast mein Interesse weiter zu verfolgen.

Thank you Benjamin, for your passion and talent to make me laugh no matter how hard (or easy) the circumstances, and making my life much more joyful.

I gratefully acknowledge funding by Balliol College and the EPSRC.

Contents

Chapter 1
Introduction

Over 100 years after its discovery, quantum physics continues to fascinate physicists and fuel modern research all over the world. While the creation of Schrödinger cat states [1], Bell inequality violations [2] and quantum teleportation [3] have turned from incredible and fanciful predictions to routine experiments in the lab, there are still plenty of questions that remain unanswered, not least of all how to interpret wave-particle duality and probability wave-functions. A tool that promises to help solving some of the big open questions is the quantum computer—a precisely controlled quantum system that can be used to simulate other quantum systems that are too complex to be simulated with a classical computer. The idea was first proposed in the early 1980's by Richard Feynman [4] and Paul Benioff [5]. Soon afterwards David Deutsch proposed the concept of a universal quantum computer, and also the first quantum algorithm that is faster than any classical equivalent [6, 7]. When in 1994 Peter Shor proposed an algorithm for quantum computers that can efficiently factorise large numbers [8], the interest of governments was sparked[1] and more funding became available. In the aftermath there was rapid progress on many fronts. On the theory side, in 1994 Cirac and Zoller proposed a mechanism for a C-NOT entangling gate with trapped ions [9], in 1995 Andrew Steane [10] and Peter Shor [11] proposed schemes for quantum error correction, in 1996 Lov Grover proposed a quantum algorithm that can efficiently search unstructured databases [12] and in 1997 David DiVincenzo compiled a list of necessary requirements for a universal quantum computer [13]. Since then, many different qubit platforms and gate mechanisms have been proposed. On the experimental side, Chris Monroe et al. realised the first C-NOT gate with trapped ions in 1995 [14], and in 1998 the first quantum algorithms were implemented in nuclear magnetic resonance (NMR) systems (Grover's search algorithm by Isaac Chuang et al. [15] and Deutsch's algorithm by Jonathan Jones and Michele Mosca [16]). Today, all three algorithms have been implemented in various systems with increasing number of qubits [17–20].

---

[1]The security of RSA, the encryption algorithm used in most public key cryptography systems, relies on the fact that factorisation of large numbers is a hard problem and can't be solved time efficiently.

© Springer Nature Switzerland AG 2020
V. M. Schäfer, *Fast Gates and Mixed-Species Entanglement with Trapped Ions*,
Springer Theses, https://doi.org/10.1007/978-3-030-40285-3_1

For characterising the suitability of a system for quantum computing, DiVincenzo's criteria [13] form a recognized standard. They state that a quantum computer requires

1. Hilbert space control, i.e. a set of well-defined addressable qubits in a scalable system.
2. State preparation
3. Low decoherence, such that the qubit's coherence time is much longer than the duration of the gate operations
4. Controlled unitary transformations, i.e. single-qubit and two-qubit operations forming a universal gate set
5. Qubit state-specific quantum measurements

While state preparation, measurement and gate operations require precisely controlled interactions with the qubit, low decoherence is best fulfilled for highly isolated systems.

A universal gate set is a set of logic operations that can be used to implement any other logic operation, and therefore any algorithm desired. For a quantum computer a popular universal gate set is the Hadamard gate H, the CNOT gate and the $\pi/8$ phase gate T, where the single-qubit gates are

$$H = \frac{1}{\sqrt{2}} \begin{pmatrix} 1 & 1 \\ 1 & -1 \end{pmatrix}, \; T = \begin{pmatrix} 1 & 0 \\ 0 & e^{i\pi/4} \end{pmatrix} \tag{1.1}$$

and the two-qubit gate is

$$\text{CNOT} = \begin{pmatrix} 1 & 0 & 0 & 0 \\ 0 & 1 & 0 & 0 \\ 0 & 0 & 0 & 1 \\ 0 & 0 & 1 & 0 \end{pmatrix} \tag{1.2}$$

We typically implement a controlled-phase gate C-PHASE as the two-qubit gate, with $U = \text{diag}(1, i, i, 1)$. It can easily be transformed into a CNOT gate using single-qubit operations: $\text{CNOT} = (I \otimes H) \cdot (T^{\dagger 2} \otimes T^{\dagger 2}) \cdot (U) \cdot (I \otimes H)$.

The most promising platforms to build a quantum computer are at present trapped ions [21], superconducting qubits [22], quantum dots [23], colour centres in diamond [24], or photonic systems [25]. While trapped ions achieve by far the longest coherence times [26, 27] and to date also the highest gate operation fidelities [26, 28, 29] and qubit connectivity [30], superconducting qubit systems have increased their coherence times and gate operation fidelities [31] in recent years and promise easier scalability.

Trapped ions were chosen early as a platform for qubits. They fit beautifully to the conditions of a small, well isolated system with precise controllability. The qubits are encoded in single atoms, fixed in space only by electric fields inside a vacuum chamber, and are therefore very well isolated from their environment. Unlike solid state qubits, which are manufactured and therefore often exhibit small differences due to manufacturing imperfections, trapped ion qubits are naturally identical to each other. Each qubit is encoded in the electronic states of an ion's valence electron, also called the internal state. The ions are spaced far enough apart that crosstalk between electrons of different ions is negligible. This high degree of isolation gives rise to very long coherence times that are typically only limited by magnetic field fluctuations. At the same time the qubits can be manipulated using lasers and microwaves, which allow very precise and flexible control. The ions' motion, also called the external state, is coupled strongly via the Coulomb interaction. This provides the interaction necessary for entangling gates.

Possibly the biggest challenge for building a quantum computer with trapped ions is scalability. Simply adding more ions to a trap is not a scalable approach, since the number of motional modes increases and the modes become closer in frequency, reducing the fidelity of entangling operations. A promising design for a scalable system with trapped ions is the quantum CCD architecture [32], where ions are split into smaller groups [33, 34] and 'shuttled' between different trap zones. Advances in trap miniaturisation and integration of optics into the trap [35] further aid scalability. By using many smaller modules in the QCCD architecture and connecting them via photonic links [36] the system can be scaled up even further.

1.1 Mixed-Species Entanglement

An important aspect in scaling up trapped ion systems is to use different species of ion. This yields several advantages.

Sympathetic cooling Because the ions' motion is used as a bus for entangling operations, it is important that the ions' temperature remains fairly constant. However, ion heating occurs in every trap and is especially pronounced in the more easily scalable surface traps. Using different kinds of ion allows cooling one species of ion, interleaved with performing logic operations on the other species. The motion of the logic species will thus be cooled 'sympathetically' by the ion of the cooling species, without its qubit state being corrupted by the cooling lasers [37].

Quantum logic spectroscopy High precision spectroscopy of atoms and molecules without closed cycling transitions for Doppler cooling or suitable methods for state preparation and readout can be difficult. Using different ion species not only allows sympathetic cooling, but can also be used to prepare and readout the state of the 'spectroscopy' ion, by coherently transferring the spin state from one ion to the other [38]. The state transfer can be achieved via the motional coupling of the ions due to the Coulomb force.

Ion-photon entanglement Using different species is also vital for connecting ion traps with photonic links. The entanglement is transferred between two ions in different trap modules by photons emitted by each ion from a superposition state. The emitted photons are entangled with the ion's qubit state via their frequency or polarisation. They are interfered on a beam splitter and coincidence detection of the photons at two detectors heralds entanglement of the two ions [39]. Because photon emission into the correct mode, photon collection, photon interference into the correct mode and photon detection are all probabilistic, this process has to be repeated many times before the desired result is achieved. It is therefore necessary to excite the ion at a high rate into the desired state from where the photon is emitted, which can be achieved with a pulsed laser. This however also increases the rate at which the excitation laser or photons emitted by the ion interact with other ions in the trap – unless they are of a different species, and therefore far off-resonant. Using at least two different species is therefore necessary to guarantee a low level of cross-talk when addressing specific ions, necessary for fast ion-photon entanglement.

Species specialisation Another important aspect to consider is that ions with zero nuclear spin are preferable for ion-photon entanglement, because the simple level structure increases the emission rate into the desired mode. Ions with hyperfine structure are however better suited for logic operations, as they have so-called 'atomic clock'-qubits that possess greatly enhanced coherence times.

There are therefore multiple incentives to use different species in the same trap and transfer entanglement between them for different operations. For the prospect of using photonic links to couple different trap modules we demonstrate the mechanism of mixed-species entanglement between $^{40}Ca^+$ and $^{43}Ca^+$. Subsequently we demonstrate entanglement of $^{88}Sr^+$ and $^{43}Ca^+$, our choice of qubit for photonic entanglement and high-fidelity logic operations respectively.

1.2 Fast Gates

An important benchmark of a qubit platform is the ratio of the duration of a gate operation to the qubit's coherence time. While trapped ions have the longest coherence times by orders of magnitude, their gate operations are also considerably longer than those in other qubit platforms. Apart from reducing errors due to spin-coherence, shorter gate times are also desirable to decrease the effects of ion heating and motional dephasing. Ion heating is a prominent source of error in surface traps, which are the most scalable version of ion traps.

Trapped-ion two-qubit gate times have until now been limited by the speed of the ions' motion, as the gate mechanisms used require that the gate dynamics are in the adiabatic regime, i.e. slow compared to the ions' motion. However this limit is not fundamental. The Coulomb interaction, which gives rise to the motional coupling, acts almost instantaneously. Hence, by devising different kinds of gate mechanisms, faster gates could be implemented.

There have been several proposals for fast gates [40–45], but their implementation had been hindered so far by experimental difficulties. The proposal by Steane et al. [44] was designed specifically to match current experimental capabilities. We have implemented this proposal, achieving entanglement gates outside of the adiabatic regime and reducing the trapped-ion two-qubit gate duration by over an order of magnitude.

1.3 Outline

Chapter 2—The Ions introduces our choice of ions, their properties and the mechanisms used to control them.

Chapter 3—Theory introduces the underlying physics of trapped ions and their interaction with laser light and microwaves. It then presents the mechanisms used for entangling ions, and what to consider when using different species. Finally it introduces the theory of fast gates and characterises errors occurring during gate operations.

Chapter 4—Apparatus describes the experimental apparatus and control system used for the work described in this thesis.

Chapter 5—Experimental Characterisation presents the results of experiments used to characterise the equipment and the performance of our apparatus. It describes how the experiment was calibrated and prepared for the two-qubit gate measurements and shows the results of measurements necessary to estimate errors during gate operations.

Chapter 6—Mixed-Species Gates presents the results of experiments entangling ions of different species, some of which were published in [46].

Chapter 7—Fast Gates presents the results of the fast gate experiments. Results in this chapter have been published in [47].

Chapter 8—Conclusion gives an overview of similar work in the field of trapped ions and compares it to the performance of other qubit platforms.

References

1. Kienzler D, et al (2016) Observation of quantum interference between separated mechanical oscillator wave packets. Phys Rev Lett 116:140402. ISSN: 1079-7114
2. Hensen B, et al (2015) Loophole-free Bell inequality violation using electron spins separated by 1.3 kilometres. Nature 526:682–686. ISSN: 0028-0836
3. Bouwmeester D, et al (1997) Experimental quantum teleportation. Nature 390:575–579. ISSN: 0028-0836
4. Feynman RP (1982) Simulating physics with computers. Int J Theor Phys 21:467–488. ISSN: 0020-7748
5. Benioff P (1982) Quantum mechanical Hamiltonian models of turing machines. J Stat Phys 29:515–546. ISSN: 0031-9007
6. Deutsch D (1985) Quantum theory, the Church-Turing principle and the universal quantum computer. Proc R Soc Lond A 400:97–117
7. Deutsch D, Jozsa R (1992) Rapid solution of problems by quantum computation. Proc R Soc Lond A 439:553–558
8. Shor PW (1994) Algorithms for quantum computation: discrete logarithms and factoring. In: Proceedings of 35th Annual Symposium on Foundations of Computer Science. pp 124–134. ISSN: 0272-5428

9. Cirac JI, Zoller P (1995) Quantum computations with cold trapped ions. Phys Rev Lett 74:4091–4094. ISSN: 0031-9007
10. Steane AM (1996) Error correcting codes in quantum theory. Phys Rev Lett 77:793–797. ISSN: 1079-7114
11. Shor PW (1995) Scheme for reducing decoherence in quantum computer memory. Phys Rev 52:2493–2496. ISSN: 1050-2947
12. Grover, LK (1996) A fast quantum mechanical algorithm for database search. In: 28th Annual ACM Symposium on the Theory of Computing. pp 212–219. ISBN: 0897917855
13. DiVincenzo DP, Loss D (1998) Quantum information is physical. Superlattices Microstruct 23:419–432. ISSN: 0749-6036
14. Monroe C, Meekhof DM, King BE, Itano WM, Wineland DJ (1995) Demonstration of a fundamental quantum logic gate. Phys Rev Lett 75:4714–4717. ISSN: 0031-9007
15. Chuang IL, Gershenfeld N, Kubinec M (1998) Experimental implementation of fast quantum searching. Phys Rev Lett 80:3408–3411. ISSN: 1079-7114
16. Jones JA, Mosca M (1998) Implementation of a quantum algorithm on a nuclear magnetic resonance quantum computer. J Chem Phys 109:1648. ISSN: 0021-9606
17. Lanyon BP, et al (2007) Experimental demonstration of a compiled version of Shor's algorithm with quantum entanglement. Phys Rev Lett 99:250505. ISSN: 0031-9007
18. DiCarlo L, et al (2009) Demonstration of two-qubit algorithms with a superconducting quantum processor. Nature 460:240–244. ISSN: 0028-0836
19. Monz T, et al (2016) Realization of a scalable Shor algorithm. Science 351:1068–1070. ISSN: 0036-8075
20. Figgatt C, et al (2017) Complete 3-Qubit Grover search on a programmable quantum computer. Nat Commun 8. ISSN: 2041-1723
21. Häffner H, Roos CF, Blatt R (2008) Quantum computing with trapped ions. Phys Rep 469:155–203. ISSN: 0370-1573
22. Wendin G (2017) Quantum information processing with superconducting circuits: a review. Reports on Progress in Physics 80, 106001. ISSN: 0034-4885
23. Kloeffel C, Loss D (2013) Prospects for spin-based quantum computing in quantum dots. Annu Rev Condens Matter Phys 4:51–81. ISSN: 1947-5454
24. Childress L, Hanson R (2013) Diamond NV centers for quantum computing and quantum networks. MRS Bull 38:134–138. ISSN: 0883-7694
25. Kok P, et al (2007) Linear optical quantum computing with photonic qubits. Rev Mod Phys 79:135–174. ISSN: 0034-6861
26. Harty TP, et al (2014) High-fidelity preparation, gates, memory, and readout of a trapped-ion quantum bit. Phys Rev Lett 113:220501. ISSN: 0031-9007
27. Wang Y, et al (2017) Single-qubit quantum memory exceeding ten-minute coherence time. Nat Photonics 11:646–650. ISSN: 0028-0836
28. Ballance CJ, Harty TP, Linke NM, Sepiol MA, Lucas DM, High-fidelity quantum logic gates using trapped-ion hyperfine qubits. Phys Rev Lett 117:060504. ISSN: 1079-7114
29. Gaebler JP, et al (2016) High-fidelity universal gate set for 9Be+ ion qubits. Phys Rev Lett 117:060505. ISSN: 1079-7114
30. Debnath S, et al (2016) Demonstration of a small programmable quantum computer with atomic qubits. Nature 536:63–66. ISSN: 1476-4687
31. Barends R, et al (2014) Superconducting quantum circuits at the surface code threshold for fault tolerance. Nature 508:500–503. ISSN: 0028-0836
32. Kielpinski D, Monroe C, Wineland DJ (2002) Architecture for a large-scale ion-trap quantum computer. Nature 417:709–711. ISSN: 0028-0836
33. Bowler R et al (2012) Coherent diabatic ion transport and separation in a multizone trap array. Phys Rev Lett 109:80502
34. Ruster T, et al (2014) Experimental realization of fast ion separation in segmented Paul traps. Phys Rev A 90:033410. ISSN: 1094-1622
35. Mehta KK, et al (2016) Integrated optical addressing of an ion qubit. Nat Nanotechnol 11:1066–1070. ISSN: 1748-3395

36. Monroe C, Kim J (2013) Scaling the ion trap quantum processor. Science 339:1164–1169. ISSN: 1095-9203
37. Kielpinski D, et al (2000) Sympathetic cooling of trapped ions for quantum logic. Phys Rev A 61:032310. ISSN: 1050-2947
38. Schmidt PO, et al (2005) Spectroscopy using quantum logic. Science 309:749–752. ISSN: 0036-8075
39. Monroe C, et al (2014) Large-scale modular quantum-computer architecture with atomic memory and photonic interconnects. Phys Rev A 89:022317. ISSN: 1050-2947
40. García-Ripoll JJ, Zoller P, Cirac JI (2003) Speed optimized two-qubit gates with laser coherent control techniques for ion trap quantum computing. Phys Rev Lett 91:157901. ISSN: 0031-9007
41. Duan L-M (2004) Scaling ion trap quantum computation through fast quantum gates. Phys Rev Lett 93:100502
42. García-Ripoll JJ, Zoller P, Cirac JI (2005) Coherent control of trapped ions using off-resonant lasers. Phys Rev A 71:062309. ISSN: 1050-2947
43. Bentley CDB, Carvalho ARR, Kielpinski D, Hope J (2013) Fast gates for ion traps by splitting laser pulses. New J Phys 15. ISSN: 1367-2630
44. Steane AM, Imreh G, Home JP, Leibfried D (2014) Pulsed force sequences for fast phase-insensitive quantum gates in trapped ions. New J Phys 16. ISSN: 1367-2630
45. Palmero M, Martinez-Garaot S, Leibfried D, Wineland DJ, Muga JG (2017) Fast phase gates with trapped ions. Phys Rev A 95:022328
46. Ballance CJ, et al (2015) Hybrid quantum logic and a test of Bell's inequality using two different atomic isotopes. Nature 528:384–386. ISSN: 0028-0836
47. Schäfer, VM, et al (2017) Fast quantum logic gates with trapped-ion qubits. Nature 555:75–78. ISSN: 0028-0836

Chapter 2
The Ions

2.1 Ion Species Criteria

The typical choice of ion for trapped ion quantum computing are earth-alkalis, volatile metals or lanthanides with a full s-orbital. Thus, after single ionization, one loosely bound electron remains in the outermost shell in a hydrogen-like configuration. Further criteria are fundamental limits for errors and technical constraints. When using several species in one trap the compatibility of the species is also relevant.

The fundamental requirements for the ion's level structure are shown in Fig. 2.1. Often there is no strong cycling transition that connects to only one qubit state. Instead a subsidiary 'shelf'-state is used. This state is dark to the cycling transition, has a long lifetime, and population from one of the qubit states can be transferred state-selectively to the shelf before fluorescence readout (called shelving). In this case the excited level for state preparation, Doppler cooling and fluorescence are typically identical. Depending on the frequency ω_0 of the qubit transition it might furthermore be desirable to have access to excited states for driving Raman transitions, see Sect. 3.2.3. This is because photons of lower frequencies (e.g. in the radio-frequency (Rf) or microwave regime, which are the typical transition frequencies of Zeeman and hyperfine qubits) have only little momentum and are therefore very inefficient in coupling to the ion's motion. For these qubits one can typically find Raman transitions with photons in the optical or UV regime, which are ideal for sideband-cooling and laser-driven gates.

2.1.1 Fundamental Limits for Errors

The level structure, state lifetimes and energy splittings affect the decay patterns and unwanted (off-resonant) excitation from other levels. Thus each operation has different fundamental error limits in different atomic species.

© Springer Nature Switzerland AG 2020
V. M. Schäfer, *Fast Gates and Mixed-Species Entanglement with Trapped Ions*,
Springer Theses, https://doi.org/10.1007/978-3-030-40285-3_2

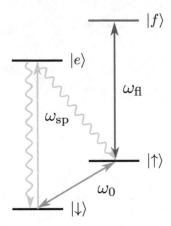

Fig. 2.1 A minimalistic ion: A trapped ion for quantum computing requires two qubit levels ($|\uparrow\rangle$, $|\downarrow\rangle$) with a long coherence time, split by the qubit frequency ω_0. It ideally has a strong cycling transition (ω_{fl}) for fluorescence readout that only couples to one qubit level, and can not decay to the other. For state preparation it is necessary to have an excited state $|e\rangle$ that can decay to both qubit levels, but where there is a laser (ω_{sp}) that excites only population out of one qubit level, selective by either polarization or frequency. Thus population can be 'pumped' into the dark state. The state-preparation and readout lasers can also be used for Doppler cooling. The lifetimes of the excited states are required to be short for fast operations and efficient cooling. The qubits are usually encoded with both states in the $S_{1/2}$ manifold (Zeeman qubits and hyperfine qubits) or one state in $S_{1/2}$ and $D_{5/2}$ each (optical qubit). The excited states are usually in the $P_{1/2}$ and/or $P_{3/2}$ manifold

Level structure Many heavier ions, e.g. Ca^+, Sr^+, Yb^+ have metastable D levels that lie lower in energy than the excited P levels. This means part of the population excited to P can decay into D, and therefore the cycling transition is no longer closed. During fluorescence detection this can be compensated for by using additional 're-pumping' lasers, that return the population in the D-states and close the cycle. However for Raman transitions D-state decay leads to a fundamental limit of the scattering error, see Sect. 3.3.

Lifetimes For optical qubits the lifetime of the metastable D state limits the coherence time of the qubit. If a D state is used as a shelf during readout, its finite lifetime limits the readout fidelity. Readout errors can be decreased by shortening the duration of the fluorescence detection. This can be achieved by reducing the background light on the detector and improving the efficiency of fluorescence detection [1].

Ground state energy splittings In our experiments state-selectivity during shelving is achieved through the frequency splitting of the two qubit levels. A larger qubit splitting is therefore desirable. Alternatively, for smaller qubit splittings, a quadrupole laser can be used for shelving. In Yb^+ the qubit splitting is large enough to provide the state-selectivity during readout, making shelving unnecessary, but again limiting readout fidelity with off-resonant excitation. Large energy splittings in the ground-state manifold are usually provided by hyperfine structure.

Hyperfine structure If an isotope's nuclear spin is not 0, the ion's level structure exhibits hyperfine levels. The combination of hyperfine structure and Zeeman splitting leads to so-called 'atomic clock'-qubits, that exhibit greatly enhanced coherence times due to robustness to magnetic field fluctuations, see Sect. 2.3. However the increased number of levels also leads to more decay paths and therefore reduces photon emission in a certain mode. This means the rate, at which photons that are useful for ion-photon (ion-photon-ion) entanglement are emitted, is reduced. Interplay of the many levels also leads to the emergence of dark resonances, complicating cooling and reducing peak fluorescence. For $I = 1/2$ the only clock qubit is at $B = 0$. This is experimentally not convenient, because an external magnetic field is necessary to define a quantisation axis and thus fix the reference frame for the laser polarisations. For high quantum logic fidelities it is therefore ideal to have a small nuclear spin $I > \frac{1}{2}$, that gives access to clock-qubits but at the same time leaves the level structure fairly simple. For ion-photon entanglement no nuclear spin is most desirable, as the simple level structure allows the largest photon emission rate in one mode.

The abundance of hyperfine levels within the qubit state manifold for atoms with a large nuclear spin also means there are transitions within the ground state manifold that are very close in frequency. This can lead to off-resonant excitation. This is especially a problem with microwaves, where it is difficult to control the radiation's polarisation. Larger Zeeman splittings usually do not help in this case, as the transitions closest in frequency stem from symmetric transitions that scale similarly with applied external magnetic field. Using composite pulse-sequences can suppress errors due to off-resonant excitation [2].

Mass Heavier ions exhibit lower motional frequencies and require larger forces to be excited into motion. The lower motional frequencies lead to lower speed-limits for adiabatic entangling gates. Larger forces mean that more laser power is required leading to larger scattering errors. However larger mass also leads to larger spin-orbit splitting, which, depending on the orbital structure, can reduce scattering into D states. In addition, lighter ions experience larger errors due to recoil caused by Rayleigh scattering [3], and are more affected by heating.

2.1.2 Technical Constraints

Although not fundamental and prospective to change with technological progress, technical constraints greatly affect the complexity, convenience and scope of experiments possible with a certain species of ion. One needs to consider the availability and performance of lasers as well as optical components and electrical/rf-control fields.

Optical transition frequency As a rule of thumb, shorter wavelengths cause more trouble. Large laser powers are more difficult to obtain, and the lasers are less reliable or more complex (e.g. requiring frequency doubling). Optical components have lower damage thresholds for UV radiation and are therefore more likely to burn. UV radiation is also more likely to charge trap electronics or burn dust onto optics. Until recently, optical fibres at beryllium UV wavelengths (313 nm) were not available [4], leading to issues with beam pointing stability. Fibre integrated AOMs are only slowly becoming available for UV wavelengths. Radiation deep in the UV can be more difficult to handle due to being carcinogenic and causing issues with free-space absorption in air. Laser diodes and optical components in the UV are also often more expensive.

Microwave transition frequency Microwave sources, amplifiers, switches and diagnostics typically become more expensive or have poorer noise specifications at higher frequencies. Acousto optic modulators (AOMs) to introduce frequency splittings for Raman lasers also become less available at higher frequencies. Phase locks at these frequencies become more technically challenging due to the difficulty in finding low noise stable reference frequency sources.

2.1.3 Compatibility of Species

To harness the advantages of different ion species it is often desirable to use multiple ion species for different tasks. To be able to exploit all benefits, such as sympathetic cooling or spectral distinction, while at the same time being able to transfer quantum information between the different species, it is important to carefully match the ion's mass and transition wavelengths. For good motional coupling it is important to have similar masses of the ions. While it is desirable to have well separated laser frequencies for addressing and sympathetic cooling, some gate mechanisms profit from similar transition frequencies, see Sect. 3.5. When using two species with hyperfine structure, it is also beneficial to find ion species that become insensitive to magnetic field fluctuations at similar magnetic fields, to increase coherence times.

2.2 Our Ions

For the work in this thesis three different ion species were used: ^{43}Ca$^+$, ^{40}Ca$^+$ and ^{88}Sr$^+$. All are alkaline-earth metals. ^{43}Ca$^+$ has a nuclear spin of $I = 7/2$, which leads to hyperfine structure. ^{40}Ca$^+$ and ^{88}Sr$^+$ have $I = 0$ and therefore do not have a hyperfine structure. An external magnetic field leads to Zeeman splitting of the individual states. There are three relevant orbitals. The ground state of the single valence electron is in the S orbital, where both of our qubit states are located. The short lived P-levels are used as excited state for Raman transitions, as well as for resonant fluorescence and Doppler cooling. The metastable long-lived D-levels are used as a shelf for readout. Both the P and D orbital exhibit fine-structure due to the spin-orbit interaction.

For quantum computing it is necessary to cool the ions close to their motional ground state, prepare a certain qubit state, perform logic operations on it and finally read out the resulting state. In the following the level structure and schemes for these operations for the ions used in this thesis are explained. If not specifically mentioned otherwise, all schemes are explained for an external magnetic field of 146 G (high field). In this work ^{88}Sr$^+$ was only used at high field, whereas ^{43}Ca$^+$ was used at low field ($B \approx 2$ G) in Sect. 6.1 and at high field in the remainder of this thesis. ^{40}Ca$^+$ was only used in Sect. 6.1 at low field. Experimental details and performance numbers will be given in Sect. 5.1.

2.2.1 Photo Ionization

For loading the ions, a current $I \sim 4$ A is driven through an oven tube filled with pellets of the desired atomic species. The current resistively heats the oven tube, which causes parts of the calcium/ strontium sample inside to evaporate. The stream of hot atoms is directed at the ion trap. Two photo ionization lasers are focussed at the centre of the trap with energies appropriate to remove one of the atom's valence electrons, see Fig. 2.2. The freshly created ions are then confined by the trap and cooled down to a crystal by Doppler cooling. A two photon process [5, 6] is used so that the first, less intense laser is resonant with a specific transition and therefore species selective, while the second, more intense laser provides the remaining energy to ionize the atom. The isotope shift between ^{43}Ca$^+$ and ^{40}Ca$^+$ for the 423 nm transition is $\delta f = 610$ MHz, which is considerably larger than the transition linewidth ($A_{12} = 2\pi \cdot 35$ MHz [7]) and therefore allows isotope-selective ionization and loading. The oven is aligned perpendicular to the photo ionisation lasers so that the excitation is insensitive to Doppler shifts. This method is used for loading at both high (146 G) and low (2 G) magnetic field [2].

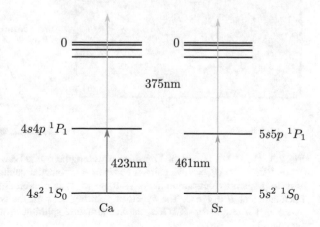

Fig. 2.2 Photo ionization: The ionization occurs in two steps. First a 461 nm (^{88}Sr) or 423 nm (^{40}Ca, ^{43}Ca) laser transfers one electron into an excited state $nsnp\ ^1P_1$. Subsequently the electron is further excited into the continuum by a 375 nm laser

2.3   Calcium 43

The level structure of ^{43}Ca$^+$ is shown in Fig. 2.3. The hyperfine and Zeeman splitting lead to a large number of ground-level states that can be chosen as qubit states. Experiments in this work were all performed on the 'stretch'-qubit $|F = 4, m_F = 4\rangle \leftrightarrow |F = 3, m_F = 3\rangle$. However, at certain magnetic fields, there exist states whose frequency difference does not in first order depend on the magnetic field, $\frac{d}{dB}(f_\uparrow - f_\downarrow) = 0$, so-called 'atomic clock'-qubits. Because magnetic field fluctuations are the main source of decoherence, these states possess a greatly increased coherence time ($T_2^* = 50$ s [8]) and are therefore ideal for quantum computing. In ^{43}Ca$^+$, the first such qubit is at $B = 0$, $|4, 0\rangle \leftrightarrow |3, 0\rangle$. However a small magnetic field is required to set the quantisation axis and lift Zeeman degeneracy. The next clock-qubit is at $B = 146.0942$ G between $|4, 0\rangle \leftrightarrow |3, 1\rangle$. The majority of work in this thesis has been performed at 146 G to have access to this special qubit.

The existence of 'clock'-qubits stems from the combination of hyperfine structure and Zeeman splitting. We can distinguish three regimes: (i) The weak field regime, where the hyperfine interaction is much larger than the Zeeman splitting. The Zeeman effect can therefore be treated as a perturbation and F and m_F are good quantum numbers. (ii) The strong field regime, where the splitting due to the external magnetic field is much larger than the hyperfine interaction, this is called the Paschen-Back regime. Here the hyperfine interaction can be treated as a perturbation and the good quantum numbers are m_I and m_J. (iii) The intermediate field regime where the Zeeman and hyperfine splitting are of comparable magnitude. Here states are in a superposition of $|F, m_F\rangle$ or $|m_J, m_I\rangle$ and the only good quantum number is

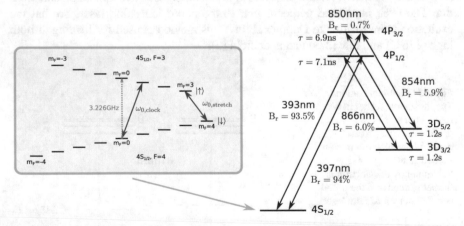

Fig. 2.3 ^{43}Ca$^+$ level structure: Transition wavelengths, excited state lifetimes and branching ratios [9] of ^{43}Ca$^+$. We choose two pairs of states as our potential qubits: $|F, m_F\rangle = |4, 4\rangle \leftrightarrow |3, 3\rangle$ is easy to prepare by optical pumping; $|4, 0\rangle \leftrightarrow |3, 1\rangle$ is first-order independent of magnetic field fluctuations at $B = 146$ G. The hyperfine splitting between the two $4S_{1/2}$ ground levels $F = 4$ and $F = 3$ is $\omega_{\text{hfs}} = 2\pi \cdot 3.2$ GHz, the fine structure splitting between $4P_{1/2}$ and $4P_{3/2}$ is $\omega_f = 2\pi \cdot 6.68$ THz

$m = m_F = m_I + m_J$. The energy shift of a state does not depend linearly on the magnetic field, but shows curvature, which leads to the emergence of clock-transitions.

Although in general the energy shifts in the intermediate field have to be calculated numerically by diagonalisation of the Hamiltonian, there exists an analytical solution for $J = \frac{1}{2}$ valid in the ground state levels of ^{43}Ca$^+$ or similar species. This solution is called the Breit-Rabi formula [10]:

$$\Delta E(F, m_F) = -\frac{\Delta E_{\text{hfs}}}{2(2I + 1)} - g_I \mu_N m_F B \pm \frac{\Delta E_{\text{hfs}}}{2} \sqrt{1 + \frac{2m_F}{I + 1/2} Bx + (Bx)^2}$$

(2.1)

$$x = \frac{g_I \mu_N + g_J \mu_B}{\Delta E_{\text{hfs}}}$$

where the nuclear spin $I = 7/2$ and the hyperfine splitting $\Delta E_{\text{hfs}} = h \cdot 3.2256082864(3)$ GHz [11]. The Landé g-factor $g_J = 2.00225664(9)$ [12] is defined over the electron's magnetic moment $\mu_J = -g_J \mu_B J$ and the nuclear g-factor $g_I = -1.315348/I$ [2] over the nuclear magnetic moment $\mu_I = g_I \mu_N I$ with nuclear magneton $\mu_N = \frac{e\hbar}{2m_p}$.

In the ground level at 146 G, although we are in the intermediate field region, the hyperfine splitting is sufficiently larger than the Zeeman splitting that F and m_F remain a natural choice to label the respective energy levels, see Fig. 2.4. However, the hyperfine splitting is smaller in higher orbitals. In 4P$_{3/2}$ and 3D$_{5/2}$ we are already in the high-field regime with appropriate quantum numbers m_I and m_J, see [2]. The qubit frequencies at 146 G are $\omega_{0,\text{stretch}} = 2\pi \cdot 2.873631432$ GHz and $\omega_{0,\text{clock}} = 2\pi \cdot 3.199941077$ GHz.

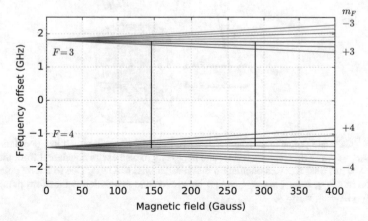

Fig. 2.4 ^{43}Ca$^+$ ground level: Applying a magnetic field lifts the degeneracy of the hyperfine levels. At $B_1 = 146.0942$ G the states $|4, 0\rangle \leftrightarrow |3, 1\rangle$ and at $B_2 = 287.7827$ G the states $|4, 1\rangle \leftrightarrow |3, 1\rangle$ form clock qubits

2.3.1   State Preparation and Readout

The qubit is prepared into $|\downarrow\rangle = 4S_{1/2} |F = 4, m_F = 4\rangle$ by optical pumping with a $\hat{\sigma}^+$ polarized 397 nm laser, see Fig. 2.5. The fidelity of the state preparation is only limited by the polarization purity of the 397 nm laser. Preparation into the clock qubit is performed by applying additional microwave pulses to transfer population along $|F = 4, m_F = 4\rangle \rightarrow |F = 3, m_F = 3\rangle \rightarrow |F = 4, m_F = 2\rangle \rightarrow |F = 3, m_F = 1\rangle$ [2]. The polarization of the microwaves is not balanced. Thus some of these transitions can only be driven with very low Rabi frequencies. Preparation into the clock-qubit is therefore very vulnerable to magnetic field noise. State preparation at low field is performed analogously.

The qubit is read out with fluorescence detection, by first transferring population from the dark state $|\downarrow\rangle$ to the shelf $3D_{5/2}$, see Fig. 2.6. The readout fidelity is limited by the lifetime of the shelf-state, and therefore by the duration of the fluorescence detection which is dictated by the net collection and detection efficiency of the imaging system and photon detector. Mis-set frequency of the 393 nm or 850 nm laser, as well as impurities in their polarisation or 854 nm laser leakage can also reduce the readout fidelity. At low field a single 850 nm laser frequency is sufficient to repump population from $3D_{3/2}$, as the energy levels are too close together to be addressed separately.

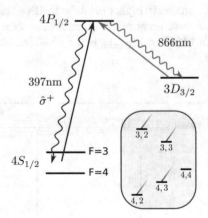

Fig. 2.5 ^{43}Ca$^+$ state preparation scheme: A $\hat{\sigma}^+$ polarized 397 nm laser excites population from $4S_{1/2}$ to $4P_{1/2}$. An EOM creates the sideband necessary to also excite population from the $F = 3$ levels. A 866 nm laser recovers population that has decayed to $3D_{3/2}$. Due to the $\hat{\sigma}^+$ polarization population in $4S_{1/2}$, $F = 4, m_F = 4$ cannot be excited and population is therefore pumped into $|\downarrow\rangle = 4S_{1/2} |F = 4, m_F = 4\rangle$

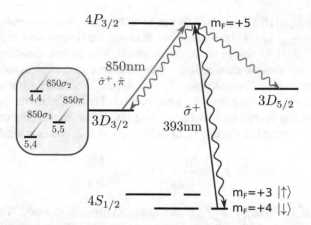

Fig. 2.6 ^{43}Ca$^+$readout scheme: A $\hat{\sigma}^+$ polarized 393 nm laser excites population from $|\downarrow\rangle =$ $4S_{1/2}^{F=4,m_F=+4}$ to $4P_{3/2}^{5,+5}$. The 3.2 GHz hyperfine-splitting suppresses excitation from the $F = 3$ levels. From $4P_{3/2}^{5,+5}$ selection rules prevent decay to any state other than $|\downarrow\rangle$in the ground state manifold. With 6% probability it can decay to the desired shelf state, $3D_{5/2}$. With 0.7% probability it can decay to $3D_{3/2}$, from where a sequence of first $\hat{\sigma}^+$and then $\hat{\pi}$ polarized 850 nm pulses transfers population back to $4P_{3/2}^{5,+5}$. A sequence of {393,850} pulses pumps all population from $|\downarrow\rangle$into the shelf. Subsequently 397 nm and 866 nm light is applied, that will cause the ion to fluoresce if there is any population left in the ground state manifold, i.e. if the ion's original state was $|\uparrow\rangle$

2.3.2 Logic Operations

Single-qubit operations are driven directly with microwaves (Sect. 3.2.2) or via the $4P$ manifold using Raman transitions (Sect. 3.2.3). In this work, two-qubit operations are performed with Raman lasers, however microwave two-qubit gates have also been demonstrated on the clock qubit in ^{43}Ca$^+$ [13].

2.3.3 Cooling

We cool ^{43}Ca$^+$with three different techniques: (i) Hot ions are cooled with Doppler cooling outside of the resolved sideband regime. The minimum temperature attainable with Doppler cooling is limited by the recoil caused by emitted photons. (ii) Dark resonance cooling [14] is a type of Doppler cooling. However it can cool below the Doppler limit by exploiting the large fluorescence gradient and low total fluorescence rate of dark resonances. (iii) Finally resolved sideband cooling is used to cool ions down to the ground state.

Doppler cooling During Doppler cooling [15] two 397 nm lasers are applied on the ion, with polarisations $\hat{\sigma}^+$ and $\hat{\pi}$. Both lasers have a sideband induced by an EOM to be resonant with population in both hyperfine manifolds. An 866 nm laser repumps population decayed to $3D_{3/2}$. The 866 nm transition is intensity broadened to cover all levels in $3D_{3/2}$. The 397 nm lasers are detuned red from resonance. Due to the Doppler effect the ions absorb mainly photons opposite to their direction of motion. The absorbed photons are then emitted again in a random direction, most likely at a higher frequency closer to the Doppler free resonance. The energy difference between the two photons is extracted from the ion's motion. The minimum temperature limit of Doppler cooling is reached once the cooling rate equals the heating rate due to the emitted photon's recoil at a detuning of $\Delta = \Gamma\sqrt{1+s}/2$, and is roughly

$$T_D \cong \frac{7\hbar\Gamma}{20k_B}\sqrt{1+s} \approx \frac{\hbar\Gamma}{2k_B} \tag{2.2}$$

with $s = 2|\Omega|^2/\Gamma^2$ [16]. For $^{40}Ca^+$ this is approximately 0.5 mK. The minimum temperature is lower for smaller Rabi frequencies. This limit however assumes a two-level system. The many levels in $^{43}Ca^+$ lead to more complex dynamics and dark resonances. These cause the achievable Doppler limit to be higher, as well as the peak fluorescence to be lower. While this type of Doppler cooling is applied during all times when the experiment is idle, during loading as well as during fluorescence detection, dark resonance cooling is used at the beginning of each experiment to cool the ion to lower temperatures.

Dark resonance cooling The minimum temperature achievable with Doppler cooling decreases with a smaller photon scattering rate, and a larger gradient of the scattering rate with laser frequency [14]. This is because a larger absolute scattering rate means an already very cold ion is more likely to absorb and emit a photon and therefore to be heated by the recoil. The large gradient means that the ion is far more likely to emit a photon with energy larger than that of the absorbed photon, rather than emitting a photon of similar or even smaller energy. At a dark resonance there is both a low total fluorescence rate and a large fluorescence gradient. By simulation of the multi-level system, parameters for optimal dark resonance cooling can be found. We use parameters similar to the ones found in [14], with small empirical adjustments to compensate for the different beam geometry. The lasers used for dark resonance cooling are the same as for Doppler cooling, with smaller laser powers and a different 866 nm detuning. Due to the narrowness of the dark resonance, the final temperature is very sensitive to the 866 nm detuning.

At low field the final temperature that can be reached with normal Doppler cooling is lower than at high field. Therefore dark resonance cooling was not necessary.

Sideband cooling For $\omega_z \gg \gamma$, with natural linewidth γ, we are in the resolved sideband regime and can address individual motional sidebands [16]. This condition is fulfilled by the typical Rabi frequencies reached by Raman transitions of $\Omega_R \approx 2\pi \cdot 130$ kHz. By tuning the Raman lasers to the red sideband, see Fig. 2.7 we can extract motional energy from the ion and cool it down to the ground state.

We distinguish between two different kinds of sideband cooling: continuous sideband cooling [17], where all lasers remain switched on simultaneously, and pulsed sideband cooling, which iterates pulses of $\{\pi_{RSB}\}$ and $\{866, 397_{\hat{\sigma}+}\}$. Continuous

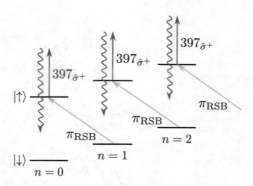

Fig. 2.7 Sideband cooling: A Raman red sideband π-pulse transfers population from $|\downarrow, n\rangle \rightarrow$ $|\uparrow, n-1\rangle$. A weak $\hat{\sigma}^+$ polarized 397 nm laser can excite population out of $|\uparrow\rangle$, but not out of $|\downarrow\rangle$, so it will eventually decay back to $|\downarrow\rangle$. A 866 nm laser repumps population from $3D_{3/2}$. Decay into the blue or red sideband is suppressed by $\eta \approx 0.18$ relative to the carrier, see Sect. 3.2.3, and therefore typically one phonon is removed in each cycle. Thus population is pumped towards $|\downarrow, 0\rangle$, from where it cannot be excited further

sideband cooling has a higher cooling rate and is therefore employed in the beginning, while pulsed sideband cooling reaches lower temperatures and is therefore used in the end. Sideband cooling at low field is performed equally to that at high field.

2.4 Strontium

The level structure of $^{88}\text{Sr}^+$is shown in Fig. 2.8. The qubit splitting is only due to Zeeman shifts and is $\omega_0 = 2\pi \cdot 408.7$ MHz at 146 G. It increases with $\Delta E = g_J \mu_B m_J B$, i.e. $\Delta f = 2.8$ MHz per Gauss.

2.4.1 State Preparation and Readout

The qubit state is prepared with frequency selective pumping with a $\hat{\sigma}^\pm$ polarised 422 nm laser, see Fig. 2.9. The limiting factor for the state preparation fidelity is the difference in the 422_l and 422_u transition frequencies.

In this work, readout is performed with a preliminary scheme, shown in Fig. 2.10. The fidelity of this scheme is rather low because population decayed into $4D_{3/2}$ is recovered only insufficiently. In future work, this readout scheme will be replaced by using a 674 nm quadrupole laser to directly transfer population into the shelf, at much higher fidelity.

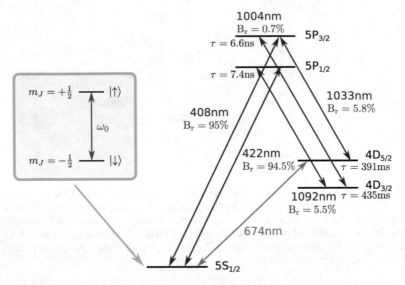

Fig. 2.8 ^{88}Sr$^+$ level structure: Transition wavelengths, excited state lifetimes and branching ratios of ^{88}Sr$^+$. Due to the lack of hyperfine structure there are only two states in the ground level and the qubit states are $|J, m_J\rangle = |1/2, -1/2\rangle \leftrightarrow |1/2, +1/2\rangle$. The fine structure splitting between $5P_{1/2}$ and $5P_{3/2}$ is $\omega_f = 2\pi \cdot 24.027$ THz. The branching ratios are calculated from the Einstein A coefficients and lifetimes tabulated in Appendix D. Due to measurement uncertainties of these values the branching ratios do not add up to 100%

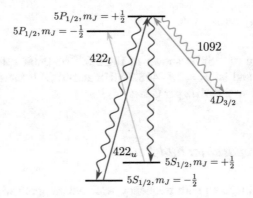

Fig. 2.9 ^{88}Sr$^+$ state preparation: A $\hat{\sigma}^\pm$ polarized 422 nm laser pumps population into one of the qubit states, while the 1092 nm laser repumps population from $4D_{3/2}$. At high field the frequency splitting between the two 422 nm transitions is 544.9 MHz, large enough to sufficiently suppress excitation from the other qubit state. Depending on which laser, 422_u or 422_l, is selected, the qubit is prepared into $|\uparrow\rangle$ or $|\downarrow\rangle$

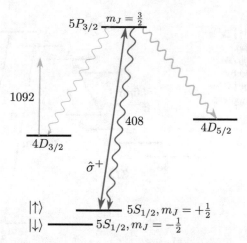

Fig. 2.10 $^{88}\text{Sr}^+$ preliminary readout scheme: A $\hat{\sigma}^+$ polarized 408 nm laser excites population from $|\uparrow\rangle = 5S^{m_J=1/2}_{J=1/2}$ to $5P^{3/2}_{3/2}$. The frequency splitting between $|\uparrow\rangle$ and $|\downarrow\rangle$ is only the Zeeman splitting at 146 G, leading to errors from off-resonant excitation. Again, $\approx 6\%$ of the population from $4P^{3/2}_{3/2}$ will decay to the shelf $3D_{\frac{5}{2}}$. Another 0.7% will decay to $3D_{\frac{3}{2}}$, from where population is recovered with 1092 nm light. It is however only recovered to a state from where it can also decay to the 'bright' qubit state $|\downarrow\rangle$, causing an error

2.4.2 Logic Operations

Single-qubit operations in $^{88}\text{Sr}^+$ are driven directly using r.f. radiation. We do not have Raman lasers with an energy splitting suitable to perform single-qubit operations in $^{88}\text{Sr}^+$, however our Raman laser wavelength is close enough to transitions in $^{88}\text{Sr}^+$, that we can use it to drive a σ_z geometric phase gate on $^{88}\text{Sr}^+$, see Sect. 6.2. In future experiments the quadrupole laser can also be used to drive entangling gates on $^{88}\text{Sr}^+$. It will also allow the performance of a mixed-species Mølmer–Sørensen gate, which depends on the qubit transition frequency, and therefore requires two sets of lasers to perform mixed species gates. The quadrupole laser will also be used to drive single-qubit operations on the ion.

2.4.3 Cooling

Doppler cooling in $^{88}\text{Sr}^+$ reaches temperatures close to the Doppler limit thanks to the simple level structure of $^{88}\text{Sr}^+$. We currently have no lasers to perform sideband cooling on $^{88}\text{Sr}^+$, instead sympathetic cooling from $^{43}\text{Ca}^+$ is used to cool $^{88}\text{Sr}^+$ below the Doppler limit to its motional ground state. In future experiments the quadrupole laser can be used to directly cool $^{88}\text{Sr}^+$ to the ground state.

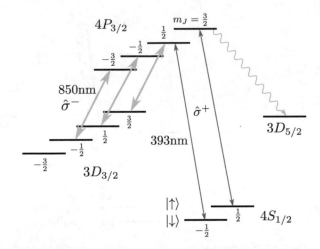

Fig. 2.11 Readout in ^{40}Ca$^+$: Population in $|\uparrow\rangle$ is shelved with a 393 nm $\hat{\sigma}^+$ polarised laser to $3D_{5/2}$. A dark resonance induced by an intense 850 nm laser ($\hat{\sigma}^-$ polarised) is used to suppress excitation from $|\downarrow\rangle$

2.5   Calcium 40

The level structure of ^{40}Ca$^+$is analogous to that of ^{88}Sr$^+$, with state lifetimes, branching ratios and transition wavelengths equal to those of ^{43}Ca$^+$, apart from an isotope shift. State preparation works through polarisation selectivity as for ^{43}Ca$^+$. At low field the frequency splittings are small enough that a single beam is sufficient and no EOM sidebands are necessary. The readout scheme is similar to that used for ^{88}Sr$^+$. However ^{40}Ca$^+$is operated at low field, and the qubit splitting is not large enough to allow state selective shelving. Instead, electromagnetically induced transparency (EIT) is used to suppress excitation out of $|\downarrow\rangle$, see Fig. 2.11.

Doppler cooling is performed equivalently to ^{88}Sr$^+$. Single-qubit operations are performed with r.f. radiation or Raman lasers. For mixed-species experiments using ^{40}Ca$^+$ and ^{43}Ca$^+$, the Raman lasers were used to sideband cool ^{40}Ca$^+$, with ^{43}Ca$^+$ being cooled sympathetically. Two-qubit entangling gates are also performed using the Raman lasers.

References

1. Noek R et al (2013) High speed, high fidelity detection of an atomic hyperfine qubit. Opt Lett 38:4735–4738. ISSN: 0146-9592
2. Harty TP (2013) High-fidelity microwave-driven quantum logic in intermediatefield 43Ca+ PhD thesis (University of Oxford, 2013)
3. Ozeri R et al (2007) Errors in trapped-ion quantum gates due to spontaneous photon scattering. Phys Rev A 75:042329. ISSN: 1050-2947
4. Colombe Y, Slichter DH, Wilson AC, Leibfried D, Wineland DJ (2014) Single-mode optical fiber for high-power, low-loss UV transmission. Opt Express 22:19783-19793. ISSN: 1094-4087

5. Gulde S et al (2001) Simple and efficient photo-ionization loading of ions for precision ion-trapping experiments. Appl Phys B: Lasers Opt 73:861-863. ISSN: 09462171
6. Lucas DM et al (2004) Isotope-selective photoionization for calcium ion trapping. Phys Rev A 69:012711. ISSN: 1050-2947
7. Kramida A, Ralchenko Y, Reader J (2017) NIST Atomic Spectra Database (ver. 5.5.1) 2017. https://physics.nist.gov/asd (2017)
8. Harty TP (2014) et al. High-fidelity preparation, gates, memory, and readout of a trapped-ion quantum bit. Phys Rev Lett 113:220501. ISSN: 0031-9007
9. Szwer D (2010) High fidelity readout and protection of a 43Ca+ trapped ion qubit PhD thesis (University of Oxford, 2010). papers://d311e016-dabd-41c6-98d5-71ce9eddf36c/Paper/p1856
10. Woodgate GK, Elementary atomic structure (Oxford University Press, London UK)
11. Arbes F, Benzing M, Gudjons T, Kurth F, Werth G (1994) Precise determination of the ground state hyperfine structure splitting of 43Ca II. Zeitschrift für Physik D 31:27–30. ISSN: 0178-7683
12. Tommaseo G et al (2003) The gJ-factor in the ground state of Ca+. Eur Phys Jo D 25:113-121. ISSN: 14346060
13. Harty TP et al (2016) High-fidelity trapped-ion quantum logic using near-field microwaves. Phys Rev Lett 117: 140501. ISSN: 10797114
14. Allcock DTC et al (2016) Dark-resonance Doppler cooling and high fluorescence in trapped Ca-43 ions at intermediate magnetic field. New J Phys 18
15. Wineland DJ, Itano WM (1979) Laser cooling of atoms. Phys Rev A 20:1521–1540. ISSN: 10502947
16. Leibfried D, Blatt R, Monroe C, Wineland D (2003) Quantum dynamics of single trapped ions. Rev Modern Phys 75:281-324. ISSN: 00346861
17. Webster S, Raman sideband cooling and coherent manipulation of trapped ions PhD thesis (University of Oxford, 2005)

Chapter 3
Theory

A trapped ion has two observables relevant for quantum computing: its internal state, which can be described as a quasi-spin, and its external state, the motion.

The quasi-spin is encoded by the energy level that the valence electron occupies. We pick two such levels as our qubit states and call them $|\uparrow\rangle$ and $|\downarrow\rangle$. For all coherent quantum computing operations the ion is prepared such that it occupies either one or a superposition of the two qubit states. The electron can interact with electromagnetic fields applied to the ion—we can use lasers, microwaves or RF radiation to manipulate the electron's state and therefore the qubit state. The spatial extent of the ion's electronic wave function is $z_0 = \sqrt{\frac{\hbar}{2M\omega_z}}$ (typically $z_0 \approx 8$ nm); this is much smaller than the separation of the ions from each other ($\approx 3\ \mu$m) or from the trap electrodes (≈ 0.5 mm). At this separation, the direct interaction between spins of different ions is very weak [1].

The ion's motion is defined by the confining electric field. In the axial direction of the trap $\hat{\mathbf{z}}$ the potential is harmonic—the motion of the ion corresponds to harmonic oscillator states. If the ion is cold enough and we have a laser with sufficiently narrow linewidth that the harmonic oscillator states are spectrally resolved, we can use this laser to excite and dampen the motion of the ion. For experiments involving the ion's motion it is desirable that the ion is close to its motional ground state. This can be achieved by using resolved sideband-cooling to extract motional energy from the ion.

If there are multiple ions in the trap, they repel each other due to their equal charge. This motional coupling between different ions is strong and basically instantaneous. A combination of the motional coupling of the ions and spin-dependent forces caused by laser radiation exciting the ions' motion can be used to create entanglement between different ions.

© Springer Nature Switzerland AG 2020
V. M. Schäfer, *Fast Gates and Mixed-Species Entanglement with Trapped Ions*,
Springer Theses, https://doi.org/10.1007/978-3-030-40285-3_3

3.1 Linear Paul Trap

The advantage of working with ions over atoms is that they can easily be trapped using electric fields due to their electric charge. Laplace's theorem $\nabla^2 \Phi = 0$ implies that no static electric field can have a minimum in three dimensions simultaneously. To confine an ion in three-dimensional space we therefore need to use a magnetic field (Penning trap [2]) or oscillating electric field (Paul trap [3]) on top of the static electric field. In our experiment we use an RF Paul trap, where four (blade or rod) electrodes arranged as a quadrupole, two of which are grounded and two of which are driven by an oscillating voltage V_{RF}, provide a ponderomotive potential as radial confinement. The two endcap electrodes are connected to DC voltages and provide a harmonic potential in the axial $\hat{z}$ direction. The resulting potential along the trap axis can be described as [4–6]

$$\Phi \simeq \frac{V_{RF}}{2} \cos(\Omega_{RF}t) \left(\frac{\kappa_z z^2}{d^2} + \frac{\kappa_x x^2 - \kappa_y y^2}{\rho^2} \right) + \alpha \frac{V_{dc}}{2} \left(\frac{2z^2 - x^2 - y^2}{d^2} \right) \quad (3.1)$$

where V_{RF} (V_{dc}) is the voltage applied to the blades (endcaps), z (x, y) is the displacement from the centre of the trap in the axial (radial) direction, ρ (d) is the distance from the trap centre to the blades (endcaps) and Ω_{RF} is the RF frequency. The parameters α, κ_x, κ_y and κ_z are geometric factors of order unity, caused by deformations of the field due to non-hyperbolic trapping electrodes and the finite size of the trap.[1] The motion of the trapped ions can be described by the Mathieu equation [6]

$$\frac{d^2 u_i}{d\xi^2} + (a_i - 2q_i \cos(2\xi)) u_i = 0 \quad (3.2)$$

with $\xi = \Omega_{RF}t/2$, $u_i \in \{x, y, z\}$, and parameters [5, 7]

$$a_x = a_y = -\frac{a_z}{2} = -\frac{4\alpha e V_{dc}}{M d^2 \Omega_{RF}^2} \quad (3.3)$$

$$q_x = -q_y = \frac{2e\kappa_i V_{RF}}{M \rho^2 \Omega_{RF}^2}, \quad q_z = \frac{2e\kappa_z V_{RF}}{M d^2 \Omega_{RF}^2}$$

For $\omega_p = \frac{e\alpha V_{RF}}{\sqrt{2}\Omega_{RF} M \rho^2} \ll \Omega_{RF}$ [8] (i.e. $q_i \ll 1$) we can approximate the trap potential as a three-dimensional harmonic well with $U(x, y, z) = \frac{1}{2}m\omega_x^2 x^2 + \frac{1}{2}m\omega_y^2 y^2 + \frac{1}{2}m\omega_z^2 z^2$ and trap frequencies $\omega_i = \frac{\Omega_{RF}}{2}\sqrt{\frac{q_i^2}{2} + a_i}$ [4, 6]. The ion's motion can then be described as a superposition of secular oscillation at ω_i in the orthogonal directions

---

[1] One often finds this potential with $\kappa_z = 0$. However the asymmetric driving of the RF blades and the finite distance to the endcap electrodes means that there is a gradient of the RF amplitude along the trap axis [5, 7]. Axial micromotion and its dependence on z can therefore not be neglected entirely. κ_z can be reduced by increasing the distance d between the trap centre and the endcaps.

x, y, z and micromotion at the frequency of the trap RF Ω_{RF}. For calculating the axial trap frequency we can assume $q_z \approx 0$ and obtain[2]

$$\omega_z \approx \sqrt{\frac{2\kappa_z e V_{dc}}{M d^2}} \propto \sqrt{\frac{V_{dc}}{M}} \tag{3.4}$$

$$\omega_r = \sqrt{\omega_p^2 - \omega_z^2/2} \propto \frac{V_{RF}}{M\Omega_{RF}}$$

For two ions of equal mass the axial mode frequencies are $\omega_{z,ip} = \omega_z$ and $\omega_{z,oop} = \sqrt{3}\omega_z$. For two ions of different masses M_1 and $M_2 = \mu M_1$ the secular mode frequencies change to [9]

$$\omega_{z,ip/oop} = \sqrt{\frac{1 + \mu \mp \sqrt{1 - \mu + \mu^2}}{\mu}}\omega_{z,1} \tag{3.5}$$

$$\omega_{x,y,ip/oop} = \sqrt{-\frac{\mu + \mu^2 - \epsilon^2(1 + \mu^2) \mp a}{2\mu^2}}\omega_{z,1}$$

where $\epsilon = \omega_p/\omega_{z,1}$, $a = \sqrt{\epsilon^4(\mu^2 - 1)^2 - 2\epsilon^2(\mu - 1)^2\mu(1 + \mu) + \mu^2(1 + (\mu - 1)\mu)}$ and $\omega_{z,1} = \omega_z(M_1)$.

The ion-spacing for ions with equal mass in a two-ion crystal can be calculated from the axial frequency with [4]

$$\delta z = \left(\frac{e^2}{2\pi\epsilon_0 M\omega_z^2}\right)^{1/3} \tag{3.6}$$

3.2 Ion-Light Interactions

Manipulation of the ions' state is performed with coherent electromagnetic fields. We will first look at the influence of these fields on the internal state of the ion and later add coupling to the external states. We begin by expressing the ion as an idealised two-level system. We roughly follow the treatment in [4, 10–15], where more detail can be found.

The internal energy of the ion H_0 is a sum of its electronic energy H_e and its motional energy H_m. We will look at a single ion with two levels $|e\rangle$, $|g\rangle$ that are split by $\hbar\omega_0$. An external magnetic field $\mathbf{B}$ sets the quantisation axis, and σ_z projects the spin of the electron on the quantisation axis with eigenvalues ± 1, such that $\sigma_z|e\rangle = |e\rangle$ and $\sigma_z|g\rangle = -|g\rangle$. The ion's motion corresponds to a harmonic oscillator along the trap axis $\hat{\mathbf{z}}$, with motional frequency ω_z. The total internal energy is then

---

[2]The scaling of the radial frequency is only approximate.

$$H_0 = H_e + H_m = \frac{\hbar\omega_0}{2}\sigma_z + \hbar\omega_z a^\dagger a \tag{3.7}$$

It is easiest to study the interaction of an ion with electromagnetic radiation by expressing the electromagnetic field in a multipole expansion [13]. To first order the interaction is described by the electric dipole operator $H_{ED} = -\mathbf{d} \cdot \mathbf{E}$. This term usually dominates, and if it is non-vanishing higher order terms can be ignored. For transitions that are electric dipole forbidden an expansion to second order is necessary. Here we obtain the magnetic dipole interaction $H_{MD} = -\boldsymbol{\mu} \cdot \mathbf{B}$ and the electric quadrupole interaction $H_{EQ} = Q\nabla \cdot \mathbf{E}$. An expansion to higher orders is not necessary for the dynamics considered in this thesis.

3.2.1 Electric Dipole Interactions

We will first study two levels $|e\rangle$ and $|g\rangle$ that are connected with a dipole allowed transition. The wavelength of the laser light is typically much larger than the size of the ion, i.e. $\mathbf{k}_l \cdot \mathbf{r}_e \ll 1$; here $\mathbf{k}_l$ is the wave vector of the laser light and $\mathbf{r}_e$ the spatial extent of the electron's wave function. We can therefore make the dipole approximation and only consider the first order term of the Hamiltonian H_{ED}. We can further assume the electric field $\mathbf{E}_l$ only depends on the position of the nucleus $\mathbf{r}$ and not on that of the electron relative to the nucleus $\mathbf{r}_e$. The interaction Hamiltonian that describes the coupling of the laser radiation to the atom then reduces to

$$H_{ED} = -\mathbf{d} \cdot \mathbf{E}_l(\mathbf{r}, t) = -q\mathbf{r}_e E_0 \epsilon_l \cos(\mathbf{k}_l \cdot \mathbf{r} - \omega_l t + \phi_l) \tag{3.8}$$

Here E_0 is the electric field amplitude, the electron's charge $q = -e$, ω_l is the laser frequency and the phase $\phi_l = \phi_0 + \phi_z$ consists of the initial optical phase of the laser oscillator ϕ_0 and the phase of the laser field at the position of the ion ϕ_z. The polarisation of the electric field is $\epsilon_l = e_0\hat{\pi} + e_+\hat{\sigma}^+ + e_-\hat{\sigma}^-$.

Due to the odd parity of the position operator $\mathbf{r}_e$ the diagonal elements of the dipole operator vanish—we can write $\mathbf{d} \cdot \epsilon_l = I(\mathbf{d} \cdot \epsilon_l)I = d_{eg}\sigma_+ + d_{eg}^*\sigma_-$ where $d_{eg} = \langle e| q\mathbf{r}_e \cdot \epsilon_l |g\rangle$, and the definitions of the Pauli matrices σ_+ and σ_- can be found in Appendix A. With the Rabi frequency $\Omega_R = -\frac{d_{eg}}{\hbar}E_0$ we obtain

$$H_I = -\frac{\hbar\Omega_R}{2}(\sigma_+ + \sigma_-)\left[e^{i(\mathbf{k}\cdot\mathbf{r}-\omega_l t+\phi_l)} + e^{-i(\mathbf{k}\cdot\mathbf{r}-\omega_l t+\phi_l)}\right] \tag{3.9}$$

This is the basic Hamiltonian describing the effect of a laser field on a two level system. In some cases we can neglect effects on the ion's motion. Then $\mathbf{k} \cdot \mathbf{r} \approx 0$ or it can be absorbed into the phase ϕ_l.

Rabi Oscillations

For studying the effects of electromagnetic radiation close to resonance of the electronic transition we first neglect effects on the motion and move to the interaction picture. We take a frame that is rotating at the ion's transition frequency ω_0. With $H_0 = \frac{\hbar\omega_0}{2}\sigma_z$ the interaction Hamiltonian $\tilde{H}_I = U^\dagger H_I U$ with $U = e^{-iH_0 t/\hbar}$ becomes

$$\tilde{H}_I = -\frac{\hbar\Omega_R}{2}\left[\sigma_+ e^{i(\delta t - \phi_l)} + \sigma_- e^{-i(\delta t - \phi_l)}\right] \tag{3.10}$$

where we have applied the rotating wave approximation (RWA) by neglecting fast oscillating terms at $\omega_0 + \omega_l$. The detuning of the laser from the ion's transition frequency is $\delta = \omega_0 - \omega_l$. On resonance $\delta = 0$ the Hamiltonian becomes time-independent and we can easily calculate the propagator

$$U = \begin{pmatrix} \cos\left(\frac{\Omega_R t}{2}\right) & i\sin\left(\frac{\Omega_R t}{2}\right)e^{-i\phi_l} \\ i\sin\left(\frac{\Omega_R t}{2}\right)e^{i\phi_l} & \cos\left(\frac{\Omega_R t}{2}\right) \end{pmatrix}$$

This propagator gives rise to Rabi oscillations that are used to transfer population coherently between two different levels and allows us to perform arbitrary rotations on the Bloch sphere. For $\phi_l = -\pi/2$ it corresponds to the rotation matrix $R_y(\theta)$, and for $\phi_l = \pi$ to $R_x(\theta)$. By adjusting the laser phase ϕ_l or, for off-resonant driving fields, by adding a time delay we can do rotations around the $\hat{\mathbf{z}}$-axis of the Bloch sphere. This Hamiltonian therefore allows us for example to implement a Hadamard gate by choosing $\theta = \Omega_R t = \pi/2$ with $H = R_y(\pi/2) \cdot R_z(\pi)$.

Light Shifts

In a real ion we will never have a perfect two-level system and radiation resonant with one transition will off-resonantly couple to other transitions. For $\delta \gg \Omega_R$ the effect on the populations will be negligible. The action of far off-resonant ion-light interaction instead becomes apparent if we go into the frame rotating with the laser frequency ω_l. We can write the electronic Hamiltonian as $H_0 = \frac{\hbar\omega_l}{2}\sigma_z + \frac{\hbar\Delta}{2}\sigma_z$, with detuning from resonance $\Delta = \omega_0 - \omega_l$. We use the notation Δ instead of δ here, to emphasise that we mainly care about this effect for large detunings. The Hamiltonian in the interaction picture then becomes

$$\tilde{H} = \tilde{H}_0 + \tilde{H}_I = \frac{\hbar\Delta}{2}\sigma_z + \frac{\hbar\Omega_R}{2}(\sigma_+ e^{-i\phi_l} + \sigma_- e^{i\phi_l}) \tag{3.11}$$

Its eigenvalues are $\lambda = \pm\frac{\hbar}{2}\sqrt{\Omega_R^2 + \Delta^2}$. For small detunings Δ this corresponds to the dressed state picture where we observe an avoided crossing at $\Delta = 0$. For large detunings $\Delta \gg \Omega_R$ we can do a Taylor expansion of the eigenvalues. Our energy then looks like

$$\tilde{H} \approx \hbar\left(\frac{\Delta}{2} + \frac{\Omega_R^2}{4\Delta}\right)\sigma_z \tag{3.12}$$

The unperturbed level at $\frac{\hbar\Delta}{2}$ is shifted by $\frac{\hbar\Omega_R^2}{4\Delta}$. This shift is called the light shift or a.c. Stark shift [11].

3.2.2 Magnetic Dipole Interactions

If the two levels $|g\rangle$ and $|e\rangle$ have the same parity, the electric dipole moment vanishes and H_{MD} and H_{EQ} become relevant. We can drive the magnetic dipole transitions using microwaves or radio-frequency (RF) radiation. The Hamiltonian then looks like

$$H_I = -\boldsymbol{\mu}\mathbf{B}(\mathbf{r}, t) = \frac{g_s\mu_B}{2}(B_x\sigma_x + B_y\sigma_y + B_z\sigma_z)\cos(\mathbf{k.r} - \omega_B t + \phi_B) \quad (3.13)$$

with magnetic moment $\boldsymbol{\mu} = -\frac{g_s\mu_B}{\hbar}\mathbf{S}$, spin operator $\mathbf{S} = \frac{\hbar}{2}\boldsymbol{\sigma}$, the electron spin-factor $g_s \approx 2$ and the Bohr magneton $\mu_B = \frac{e\hbar}{2m_e}$. For microwaves the size of the electron's wave packet is much smaller than the wavelength of the magnetic radiation and we can therefore neglect $\mathbf{k.r} \approx 0$. We can write the Hamiltonian in matrix form

$$H_I = \frac{g_s\mu_B}{2}\begin{pmatrix} B_z & B_x - iB_y \\ B_x + iB_y & -B_z \end{pmatrix}\cos(\omega_B t - \phi_B) \quad (3.14)$$

The diagonal σ_z terms cause a differential Zeeman-shift that merely shifts the energy of our two levels; we will ignore them for now. Defining $\tilde{B}_x e^{i\phi_p} = B_x - iB_y$ we can express the off-diagonal terms as $\tilde{B}_x e^{i\phi_p}\sigma_+ + \tilde{B}_x e^{-i\phi_p}\sigma_-$. After moving into the rotating frame with $H_0 = \frac{\hbar\omega_0}{2}\sigma_z$ and applying the RWA we obtain

$$\tilde{H}_I = \frac{g_s\mu_B}{4}\tilde{B}_x\left[e^{i(\phi_p+\phi_B)}e^{i\delta t}\sigma_+ + e^{-i(\phi_p+\phi_B)}e^{-i\delta t}\sigma_-\right] \quad (3.15)$$

We can see that in the RWA the Hamiltonian for an arbitrarily polarised magnetic field is equivalent to that of a field purely polarised in the $\hat{\mathbf{x}}$ direction, with an offset of ϕ_p of the initial phase. With Rabi frequency $\Omega_R = \frac{g_s\mu_B\tilde{B}_x}{2\hbar}$ this Hamiltonian has the exact same form as Eq. (3.10) and gives rise to Rabi oscillations.

3.2.3 Raman Interactions

Magnetic dipole transitions can also be driven using lasers via a stimulated Raman interaction. Instead of directly driving the transition, we have two lasers (labelled r and b) coupling $|e\rangle$ and $|g\rangle$ to a third auxiliary level $|3\rangle$. The lasers differ in frequency by the transition frequency ω_0 and are far detuned from the auxiliary level to avoid scattering photons, see Fig. 3.1. If the Raman detuning $\Delta \gg \Omega_{R_{r,b}}, \delta$ the change

Fig. 3.1 Raman transition: Population can be transferred coherently between two levels by coupling them with two laser beams that are far detuned (Δ) from a third auxiliary level. The frequency difference of the two lasers is $\omega_b - \omega_r = \omega_0 + \delta$

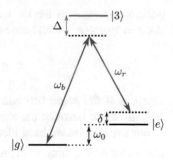

in population of the auxiliary level becomes negligible and we can adiabatically eliminate it.

A mathematically more elegant method is to apply the James-Jerke approximation [5, 16]. We start with the Hamiltonian in the frame rotating with $\omega_{g3} = \omega_b + \Delta$. With two laser-fields with frequencies ω_r and ω_b we have after the RWA:

$$\tilde{H}_I = -\frac{\hbar}{2} \Big[\Omega_{R,r} e^{i(\delta_r t - \mathbf{k}_r \cdot \mathbf{r} + \phi_{l,r})} |3\rangle\langle e| + \Omega_{R,b} e^{i(\delta_b t - \mathbf{k}_b \cdot \mathbf{r} + \phi_{l,b})} |3\rangle\langle g| + \text{h.c.} \Big] \quad (3.16)$$

where $\delta_b = \Delta$ and $\delta_r = \Delta - \delta$. Averaging over the fast dynamics evolving at $e^{-i\Delta t}$ we obtain

$$H_{\text{eff}} = \frac{\hbar\Omega_{R,b}^2}{4\Delta} (|3\rangle\langle 3| - |g\rangle\langle g|) + \frac{\hbar\Omega_{R,r}^2}{4(\Delta + \delta)} (|3\rangle\langle 3| - |e\rangle\langle e|)$$
$$- \frac{\hbar\Omega_{R,r}\Omega_{R,b}}{4\Delta_R} \Big[e^{i(\Delta\mathbf{k}\cdot\mathbf{r} + \Delta\phi_l + \delta t)} |e\rangle\langle g| + \text{h.c.} \Big] \quad (3.17)$$

where $\frac{1}{\Delta_R} = \frac{1}{2}\left(\frac{1}{\Delta} + \frac{1}{\Delta+\delta}\right) \approx \frac{1}{\Delta}$ is the effective Raman detuning, $\Delta\mathbf{k} = \mathbf{k}_r - \mathbf{k}_b$ the difference wave-vector and $\Delta\phi_l = \phi_{l,r} - \phi_{l,b}$ the relative phase of the two Raman fields. The first two terms are light shifts and can be included in the detuning δ. If we move into the frame rotating with ω_0, apply the RWA and use the effective Rabi frequency $\Omega_R = \frac{\Omega_{R,b}\Omega_{R,r}}{2\Delta}$ this Hamiltonian is again equivalent to Eq. (3.10).

Coupling to the Ion's Motion

In contrast to microwaves, for Raman transitions the term $\mathbf{k}.\mathbf{r}$ is not negligible, because the wavelength of the laser light is much shorter—in $^{43}\text{Ca}^+$ $\lambda_{\text{microwave}} \approx$ 10 cm and $\lambda_{\text{Raman}} \approx 400$ nm. This means the momentum of the photons is large enough that we can use Raman transitions to influence the motion of an ion.

Let us only consider motion along the axial direction of the trap $\hat{\mathbf{z}}$.[3] Since the motion along the trap axis corresponds to a harmonic oscillator we can express the

---

[3]The work in this thesis was performed on axial modes only. Motion along the radial direction can be derived analogously.

position operator of the nucleus $\mathbf{r} = z_0(a + a^\dagger)\hat{\mathbf{z}}$. We can then write the product $\Delta\mathbf{k}.\mathbf{r} = \eta(a + a^\dagger)$ with Lamb-Dicke parameter

$$\eta = 2\,kz_0 \sin(\frac{\alpha}{2}) \cos(\beta) \tag{3.18}$$

where α is the angle between the two Raman beam wave-vectors $\mathbf{k}_r$ and $\mathbf{k}_b$ and β is the angle between the trap axis $\hat{\mathbf{z}}$ and the Raman difference wave-vector $\Delta\mathbf{k}$. In our experiment we have ideally $\alpha = \pi/2$, $\beta = 0$ and we obtain the Lamb-Dicke parameter $\eta = \sqrt{2}kz_0$, with ground state wave-function spread $z_0 = \sqrt{\frac{\hbar}{2M\omega_z}}$.

The Lamb-Dicke parameter can also be expressed in terms of the square-root of the ratio of the photon's recoil energy $E_R = \frac{(\hbar k)^2}{2M}$ on the atom to the energy of a motional quantum (phonon) $E_m = \hbar\omega_z$, $\eta^2 = \frac{E_R}{E_m}$. A larger Lamb-Dicke parameter therefore means the emission/absorption of a single photon has a stronger effect on the ion's motion.

We begin with the Raman interaction Hamiltonian (3.17), with the light shifts included in δ

$$H_{\text{eff}} = -\frac{\hbar\Omega_R}{2} \left[e^{i[\eta(a+a^\dagger)+\Delta\phi_l-\delta t]}\sigma_+ + \text{h.c.} \right] \tag{3.19}$$

and by moving into the interaction picture with respect to $H_m = \hbar\omega_z a^\dagger a$ we obtain

$$\tilde{H}_{\text{eff}} = -\frac{\hbar\Omega_R}{2} \left[e^{i[\eta(ae^{-i\omega_z t}+a^\dagger e^{i\omega_z t})+\Delta\phi_l-\delta t]}\sigma_+ + \text{h.c.} \right] \tag{3.20}$$

Lamb-Dicke regime: We can simplify this Hamiltonian if the ion's wave-packet is very small compared to the laser wavelength, i.e. if the ion experiences a constant phase and amplitude of the laser light over the entire extent of its wave-packet: $\langle \Psi_{\text{motion}} | k^2 z^2 | \Psi_{\text{motion}} \rangle \ll 1$ or $\eta^2(2n + 1) \ll 1$. When this condition is fulfilled we are in the Lamb-Dicke regime and can neglect higher-order terms in the Taylor expansion of the exponential

$$\tilde{H}_{\text{eff}} = -\frac{\hbar\Omega_R}{2} \left[\left(1 + i\eta \left[ae^{-i\omega_z t} + a^\dagger e^{i\omega_z t}\right] + \mathcal{O}(\eta^2)\right) e^{i(\Delta\phi_l - \delta t)}\sigma_+ + \text{h.c.} \right] \tag{3.21}$$

It is a necessary condition for the Lamb-Dicke regime that $\eta \ll 1$. The impact of a single photon can therefore maximally add or remove one phonon. This leads to three interesting cases, see Fig. 3.2: (i) $\delta = 0$, the 'carrier' transition, no phonons are created or destroyed; (ii) $\delta = -\omega_z$, the 'red sideband' transition (RSB), each photon absorption 'removes' one phonon; and (iii) $\delta = +\omega_z$, the 'blue sideband' transition (BSB), each photon absorption 'adds' one phonon.

Carrier transitions: For $\delta = 0$ the resonant term $(1 \cdot e^{i\Delta\phi_l}\sigma_+ + \text{h.c.})$ dominates. The other terms are fast-oscillating and suppressed by $\eta \ll 1$ and can therefore in general be neglected. Thus our dynamics can be described to a good approximation by

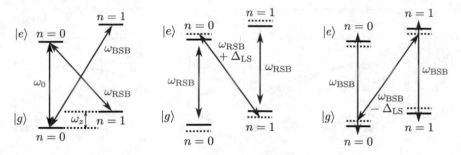

Fig. 3.2 Sideband transitions: Left: Carrier, red and blue sideband transitions in the Lamb-Dicke approximation. Middle and right: Lightshifts on the sidebands due to coupling to the carrier transition. Typically $\omega_0 \approx 2\pi \cdot 3\,\mathrm{GHz} \gg \omega_z \approx 2\pi \cdot 2\,\mathrm{MHz} \gg \Omega_{\mathrm{R,carrier}} \approx 2\pi \cdot 100\,\mathrm{kHz} \gg \Delta_{\mathrm{LS}} \approx 2\pi \cdot 5\,\mathrm{kHz}$

the Rabi oscillation Hamiltonian from Eq. (3.10), coupling the levels $|g, n\rangle \leftrightarrow |e, n\rangle$. The Rabi frequency is independent of n within the Lamb-Dicke approximation.

Red sideband transitions: For $\delta = -\omega_z$ the Hamiltonian becomes

$$\tilde{H}_{\mathrm{RSB}} = -\frac{\hbar\Omega_{\mathrm{R}}}{2} \left(e^{i(\Delta\phi_l + \omega_z t)}\sigma_+ + i\eta \left[a + a^\dagger e^{2i\omega_z t} \right] e^{i\Delta\phi_l}\sigma_+ + \mathrm{h.c.} \right) \qquad (3.22)$$

Although off-resonant, the first carrier term can not be entirely neglected as it is enhanced by a factor of $1/\eta$. For $\omega_z \gg \Omega_{\mathrm{R}}$ it leads to a light-shift that increases the transition frequency. The far off-resonant term $a^\dagger e^{2i\omega_z t}$ can however be neglected and our dynamics become

$$\tilde{H}_{\mathrm{RSB}} = -\frac{\hbar\Omega_{\mathrm{R}}}{2} i\eta \left(a e^{i\Delta\phi_l}\sigma_+ - a^\dagger e^{-i\Delta\phi_l}\sigma_- \right) \qquad (3.23)$$

This corresponds to the Jaynes-Cummings Hamiltonian known from cavity QED, where the photon field inside the cavity has been replaced by phonons. It drives Rabi oscillations between the levels $|e, n\rangle \leftrightarrow |g, n+1\rangle$ and in the process leads to entanglement between the motional and internal degrees of freedom of the ion. We can calculate the modified Rabi-frequency by

$$\sum_{n,n'=0}^{\infty} \left\langle n \left| \tilde{H}_{\mathrm{RSB}} \right| n' \right\rangle = -i\frac{\hbar}{2}\eta\sqrt{n+1}\,\Omega_{\mathrm{R}} \left(e^{\Delta\phi_l}\sigma_+ + e^{-\Delta\phi_l}\sigma_- \right) \qquad (3.24)$$

where $n = n_< = n_{|e\rangle}$. Hence $\Omega_{\mathrm{R}_{n,n'},\mathrm{RSB}} = i\eta\sqrt{n_< + 1}\,\Omega_{\mathrm{R}}$.

Blue sideband transitions: Analogously for $\delta = +\omega_z$ we obtain the anti-Jaynes-Cummings Hamiltonian

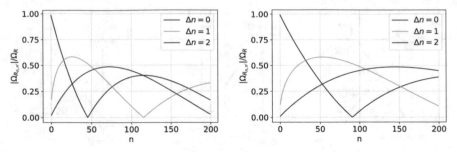

Fig. 3.3 Laguerre polynomials: Rabi frequency of the carrier, first and second sideband outside of the Lamb-Dicke approximation. Left: $\eta = 0.178$ (single $^{43}\text{Ca}^+$ ion at $f_z = 1.86\,\text{MHz}$) Right: $\eta = 0.126$ (ip mode of two $^{43}\text{Ca}^+$ ions at same trap frequency)

$$\tilde{H}_{\text{BSB}} = -\frac{\hbar\Omega_{\text{R}}}{2} i\eta \left(a^\dagger e^{i\Delta\phi_l} \sigma_+ - a e^{-i\Delta\phi_l} \sigma_- \right) \tag{3.25}$$

It couples the two levels $|e, n+1\rangle \leftrightarrow |g, n\rangle$ with effective Rabi-frequency $\Omega_{\text{R}_{n,n'},\text{BSB}} = i\eta\sqrt{n_>}\Omega_{\text{R}}$. The carrier light shift reduces the transition frequency.

Beyond the Lamb-Dicke regime: The condition that $\eta \ll 1$ is typically fulfilled fairly well for trapped ions. However the more stringent condition $\eta^2(2n+1) \ll 1$ can easily be violated if the ions are hot or have a large displacement in phase-space. In this case the Lamb-Dicke approximation is no longer valid and we have to take higher orders in η into account. The full Rabi frequency is [4]

$$\Omega_{\text{R}_{n,n'}} = \Omega_{\text{R}} \left| \langle n' | e^{i\eta(a+a^\dagger)} | n \rangle \right|$$

$$= \Omega_{\text{R}} e^{-\eta^2/2} \sqrt{\frac{n_<!}{n_>!}} \eta^{|n'-n|} \mathscr{L}_{n_<}^{|n'-n|}(\eta^2) \tag{3.26}$$

with Laguerre polynomials $\mathscr{L}_n^\alpha(X) = \sum_{m=0}^n (-1)^m \frac{n+\alpha}{n-m} \frac{X^m}{m!}$. Therefore, compared to within the Lamb-Dicke regime, the carrier Rabi frequency decreases for higher n. The sideband Rabi frequency increases more slowly than its initial $\sqrt{n}$-dependency and eventually also decreases, see Fig. 3.3.

3.2.4 Scattering

Another effect of the interaction between a trapped ion and electromagnetic radiation is scattering. A photon from the driving laser field is absorbed by the ion, transferring its valence electron into an excited state. The excited state then decays via spontaneous emission and a photon is emitted into a random direction. The total scattering rate is given by the Kramers-Heisenberg formula [13, 17]:

$$\Gamma_{i,f} = \frac{g^2\gamma}{4} \sum_k \left| \sum_J \frac{a_{i\to f}^{(J)}(k)}{\Delta_J} \right|^2$$

$$a_{i\to f}^{(J)}(k) = \frac{1}{\mu^2} \sum_q \sum_{e\in J} \langle f | \mathbf{d}\cdot\hat{\sigma}^q | e \rangle \langle e | \mathbf{d}\cdot e_k\hat{\sigma}^k | i \rangle \tag{3.27}$$

$$\Gamma_{i,\text{total}} = \sum_f \Gamma_{i,f} = \Gamma_{\text{Raman}} + \Gamma_{\text{Rayleigh}}$$

Here $g = E\mu/\hbar$ is the laser coupling, $E = \sqrt{2I/c\epsilon_0}$ the electric field amplitude, γ the radiative linewidth of the excited state,[4] $\mu = |\langle e | \mathbf{d}\hat{\sigma} | g \rangle|$ the electric dipole moment of the cycling transition, $\epsilon_l = \sum_k e_k\hat{\sigma}^k$ the polarisation of the incoming light, $\hat{\sigma}^q = (\hat{\sigma}^+, \hat{\sigma}^-, \hat{\pi})$ the polarisation of the emitted photon, $|i\rangle$ the initial state, $|f\rangle$ the final state and $|e\rangle$ an intermediate excited state within the excited J-manifold. The detuning of the incoming laser light to the excited state manifold is Δ_J, and for a Gaussian beam of waist radius w_0 the peak laser intensity at its centre is $I = \frac{2P}{\pi w_0^2}$.

The total scattering can be split into two parts: (i) Inelastic Raman scattering where $i \neq f$. The polarisation and frequency of the emitted photon depend on, and are entangled with, the ion's internal state. This acts as a measurement of the qubit state. Each Raman scattering event therefore leads to total loss of the ion's coherence. (ii) Elastic Rayleigh scattering where $i = f$. Here the photon does not carry any information about the internal state of the ion [17], which would decohere the ion's internal qubit state. However each elastic scattering event produces a phase shift that depends on the excitation path of the electron. If the scattering rates of the two qubit states are not equal, this phase shift leads to decoherence of the qubit. This elastic dephasing rate can be quantified with [5, 18]

$$\Gamma_{el} = \frac{g^2\gamma}{4} \sum_k \left| \sum_J \frac{a_{\uparrow\to\uparrow}^{(J)}(k) - a_{\downarrow\to\downarrow}^{(J)}(k)}{\Delta_J} \right|^2 \tag{3.28}$$

A further effect of photon scattering is the momentum kick the ion experiences due to absorption and emission of a photon. This leads to decoherence of the motional state [19].

3.3 Transition Elements

We can quantify the Rabi frequencies and scattering rates with the quantum numbers of the specific transitions in ^{43}Ca$^+$ and ^{88}Sr$^+$, listed in Sect. 2.3. The electric dipole transition matrix elements $d_{ge} = \langle g | \mathbf{d}_e\epsilon_l | e \rangle$ can be calculated using the Wigner-Eckart theorem. Following [15] we obtain for a fine-structure transition

---

[4]We note that technically γ also depends on J. However for lighter atoms with small spin-orbit splitting the dependence is not very strong and the mathematical treatment is more elegant neglecting the J-dependence.

$$\langle g|\, \mathbf{d}_e \epsilon_l\, |e\rangle = \sum_q e_q \left\langle L, J, m_J \,\middle|\, d_q \,\middle|\, L', J', m_J' \right\rangle \tag{3.29}$$

$$= \sum_q e_q \left\langle L\, ||\mathbf{d}||\, L'\right\rangle (-1)^{L+S-m_J} \sqrt{(2J'+1)(2J+1)(2L+1)} \tag{3.30}$$

$$\begin{Bmatrix} L & L' & 1 \\ J' & J & S \end{Bmatrix} \begin{pmatrix} J' & 1 & J \\ m_J' & -q & -m_J \end{pmatrix}$$

where L, J, m_J are the quantum numbers of the state lower in energy, L', J', m_J' are the quantum numbers of the state higher in energy, e_q are the polarisation components of the absorbed or emitted photon with $q = m_J' - m_J$. {} corresponds to the Wigner-6j symbol and () is the Wigner-3j symbol. We can infer the reduced matrix element $\langle L\, ||\mathbf{d}||\, L'\rangle$ from the Einstein A coefficient with

$$\langle L_g\, ||\mathbf{d}||\, L_e\rangle = \sqrt{\frac{3\pi\epsilon_0 \hbar c^3}{\omega_0}} A_{eg} \frac{(-1)^{J_e+L+S+1}}{\sqrt{(2J+1)(2L+1)}} \Bigg/ \begin{Bmatrix} L & L_e & 1 \\ J_e & J & S \end{Bmatrix}$$

This gives $\langle L_g\, ||\mathbf{d}||\, L_e\rangle_{\mathrm{Ca}} = 3.4930 e a_0$ for Ca^+ and $\langle L_g\, ||\mathbf{d}||\, L_e\rangle_{\mathrm{Sr}} = 3.7727 e a_0$ for Sr^+.

For hyperfine transitions we obtain [15]

$$\langle g|\, \mathbf{d}_e \epsilon_l\, |e\rangle = \sum_q e_q \left\langle L, J, F, m_F \,\middle|\, d_q \,\middle|\, L', J', F', m_F' \right\rangle \tag{3.31}$$

$$= \sum_q e_q \left\langle L\, ||\mathbf{d}||\, L'\right\rangle \sqrt{(2F'+1)(2F+1)(2J'+1)(2J+1)(2L+1)}$$

$$(-1)^{I+J+J'+L+S+1-m_F} \begin{Bmatrix} J & J' & 1 \\ F' & F & I \end{Bmatrix} \begin{Bmatrix} L & L' & 1 \\ J' & J & S \end{Bmatrix} \begin{pmatrix} F' & 1 & F \\ m_F' & -q & -m_F \end{pmatrix}$$

We normalise matrix elements as $\langle g|\, \mathbf{d}_e \epsilon_l\, |e\rangle_n = \langle g|\, \mathbf{d}_e \epsilon_l\, |e\rangle /\mu$ where μ is the matrix element of a cycling transition (and therefore the largest matrix element). For Calcium we choose in agreement with [5]

$$\mu_{\mathrm{Ca}} = \left\langle L=0, J=1/2, F=4, m_F=4 \,\middle|\, d_{q=+1} \,\middle|\, L'=1, J'=3/2, F=5, m_F'=5 \right\rangle_{43\mathrm{Ca}^+}$$

$$= \frac{1}{\sqrt{3}} \left\langle L\, ||\mathbf{d}||\, L'\right\rangle_{\mathrm{Ca}} = 2.0167 e a_0 \tag{3.32}$$

For Strontium we obtain

$$\mu_{\mathrm{Sr}} = \left\langle L=0, J=1/2, m_J=-1/2 \,\middle|\, d_{q=+1} \,\middle|\, L'=1, J'=3/2, m_J'=3/2 \right\rangle_{88\mathrm{Sr}^+}$$

$$= \frac{1}{\sqrt{3}} \left\langle L\, ||\mathbf{d}||\, L'\right\rangle_{\mathrm{Sr}} = 2.1782 e a_0 \tag{3.33}$$

The normalisation with μ gives an easily readable representation of the matrix elements that can be computed without the measured quantities included in $\langle L \,||\mathbf{d}||\, L' \rangle$. The symmetry of the dipole operator with respect to the atomic orbitals is represented by the Wigner-3j and -6j symbols: they vanish for dipole-forbidden transitions. For calculating the Rabi-frequency of Raman transitions between the levels $|\uparrow\rangle$ and $|\downarrow\rangle$ we can therefore simply take the sum of transition elements over all intermediate states:

$$\Omega_{\text{Raman},\uparrow\downarrow} = \frac{\Omega_{R,r}\Omega_{R,b}}{2\Delta} = \sum_{J=\frac{1}{2},\frac{3}{2}} \sum_{e \in J} \frac{\langle \downarrow |\mathbf{d}_e \epsilon_{l,b}| e \rangle \langle e |\mathbf{d}_e \epsilon_{l,r}| \uparrow \rangle}{2\Delta_J \mu^2} g_r g_b \qquad (3.34)$$

Here $g_{r/b} = E_{r/b}\mu/\hbar$ are the coupling constants of the two involved laser beams, $\epsilon_{l,b/r}$ their respective polarisations, and Δ_J the Raman detuning to the corresponding P_J manifold. Analogously we can calculate the light shift caused by two Raman beams on a single level $|\uparrow\rangle$ with

$$\Omega_{\text{Raman},\uparrow} = \sum_{J=\frac{1}{2},\frac{3}{2}} \sum_{e \in J} \frac{\langle \uparrow |\mathbf{d}_e \epsilon_{l,b}| e \rangle \langle e |\mathbf{d}_e \epsilon_{l,r}| \uparrow \rangle}{2\Delta_J \mu^2} g_r g_b \qquad (3.35)$$

This will give rise to the driving force for two-qubit entangling gates, see Sect. 3.4.

If the light shift of a laser beam differs for two qubit levels, the difference will change the qubit frequency ω_0. We can calculate this differential light shift in frequency units via

$$\Delta_{\text{LS}} = \frac{\Omega_{R,\uparrow}^2}{4\Delta} - \frac{\Omega_{R,\downarrow}^2}{4\Delta} \qquad (3.36)$$

$$= \sum_{J=\frac{1}{2},\frac{3}{2}} \sum_{e \in J} \frac{\langle \uparrow |\mathbf{d}_e \epsilon_l| e \rangle \langle e |\mathbf{d}_e \epsilon_l| \uparrow \rangle - \langle \downarrow |\mathbf{d}_e \epsilon_l| e \rangle \langle e |\mathbf{d}_e \epsilon_l| \downarrow \rangle}{4\Delta_J \mu^2} g^2$$

Typically the laser polarisations are chosen such that the differential light shift is zero. However it can be useful to measure the differential light shift with purely $\hat{\sigma}^+$-polarised light to calibrate laser intensities.

Scattering

We can calculate the scattering rates in a similar fashion using Eqs. 3.27 and 3.28. A summary of all relevant Rabi frequencies, scattering rates and light shifts can be found in Table 3.1. Formulas for arbitrary polarisations can be found in the Appendix B.

We notice that for large Δ the Raman scattering rate is suppressed relative to the total scattering rate. This is because during Raman scattering the spin of the electron is flipped. The electric dipole radiation of the Raman photons does not couple to the electron's spin, instead the spin-flip is mediated by the spin-orbit interaction [19]. The contribution to this coupling of the two $J = \frac{1}{2}, \frac{3}{2}$ levels is of opposite sign. For $\Delta \gg \omega_f$ these terms therefore cancel which leads to the observed suppression of

Table 3.1 Raman laser interactions: Rabi frequencies, light shifts and scattering rates of the stretch qubit in ^{43}Ca$^+$ and the Zeeman qubit in ^{88}Sr$^+$, in agreement with [5]. We assume we are in the low field regime and thus the quantum numbers of the states are $|\uparrow\rangle = |J = \frac{1}{2}, F = 3, m_F = 3\rangle$ and $|\downarrow\rangle = |J = \frac{1}{2}, F = 4, m_F = 4\rangle$ for ^{43}Ca$^+$ and $|\uparrow\rangle = |J = \frac{1}{2}, m_J = \frac{1}{2}\rangle$ and $|\downarrow\rangle = |J = \frac{1}{2}, m_J = \frac{-1}{2}\rangle$ for ^{88}Sr$^+$. α is the correction factor for ^{43}Ca$^+$ at intermediate field. The polarisations correspond to the ones used in this work, i.e. $\{\hat{\sigma}^\pm\hat{\pi}\}$ for Rabi flops, $\{\hat{\sigma}^\pm\hat{\sigma}^\mp\}$ for the light shift forces, $\{\hat{\sigma}^\pm\}$ or $\{\hat{\sigma}^\mp\}$ for all scattering rates and pure $\{\hat{\sigma}^\pm\}$, $\{\hat{\sigma}^\mp\}$ or $\{\hat{\pi}\}$ for the differential light shift. The total scattering rate is $\Gamma_{tot} = (\Gamma_{\uparrow,\text{total}} + \Gamma_{\downarrow,\text{total}})/2$

	^{43}Ca$^+$	^{88}Sr$^+$	α
$\Omega_{\text{Raman},\uparrow\downarrow}$	$\frac{\sqrt{14}g_r g_b}{24}\frac{\omega_f}{\Delta(\Delta-\omega_f)}$	$-\frac{g_r g_b}{6}\frac{\omega_f}{\Delta(\Delta-\omega_f)}$	0.98
$\Omega_{\text{Raman},\downarrow}$	$\frac{g_r g_b}{6}\frac{\omega_f}{\Delta(\Delta-\omega_f)}$	$-\frac{g_r g_b}{6}\frac{\omega_f}{\Delta(\Delta-\omega_f)}$	1.00
$\Omega_{\text{Raman},\uparrow}$	$-\frac{g_r g_b}{8}\frac{\omega_f}{\Delta(\Delta-\omega_f)}$	$\frac{g_r g_b}{6}\frac{\omega_f}{\Delta(\Delta-\omega_f)}$	0.92
Δ_{LS}	0	0	–
Γ_{tot}	$\frac{2\Delta^2+(\Delta-\omega_f)^2}{12\Delta^2(\Delta-\omega_f)^2}\gamma g^2$	$\frac{2\Delta^2+(\Delta-\omega_f)^2}{12\Delta^2(\Delta-\omega_f)^2}\gamma g^2$	1.00
$\Gamma_{\text{Raman},\uparrow}$	$\frac{23\gamma g^2}{576}\frac{\omega_f^2}{\Delta^2(\Delta-\omega_f)^2}$	$\frac{\gamma g^2}{36}\frac{\omega_f^2}{\Delta^2(\Delta-\omega_f)^2}$	1.06
$\Gamma_{\text{Raman},\downarrow}$	$\frac{\gamma g^2}{36}\frac{\omega_f^2}{\Delta^2(\Delta-\omega_f)^2}$	$\frac{\gamma g^2}{36}\frac{\omega_f^2}{\Delta^2(\Delta-\omega_f)^2}$	1.00
Γ_{el}	$\frac{49\gamma g^2}{576}\frac{\omega_f^2}{\Delta^2(\Delta-\omega_f)^2}$	$\frac{\gamma g^2}{9}\frac{\omega_f^2}{\Delta^2(\Delta-\omega_f)^2}$	0.93

Raman scattering. The ratio of both $\Gamma_{el}/\Omega_{\text{Raman},\uparrow\downarrow}$ and $\Gamma_{\text{Raman}}/\Omega_{\text{Raman},\uparrow\downarrow}$ goes to 0 for $\Delta \to \pm\infty$. This means we can arbitrarily suppress the errors due to Raman scattering and elastic dephasing by increasing the laser detuning and power while maintaining constant Rabi frequency. However this is not the case for the total scattering rate: $\lim_{\Delta\to\infty} \Gamma_{\text{tot}}/\Omega_{\text{Raman},\uparrow\downarrow} = \text{const} \neq 0$. The scattering into the D states is proportional to Γ_{tot} with $\Gamma_D = B_{r,D}\Gamma_{\text{tot}}$. This is therefore a fundamental error that cannot be suppressed further and limits the possible fidelity for Raman laser gates in Ca$^+$. For two qubit entangling gates this error is $\epsilon_{D,\infty} \approx 1 \cdot 10^{-4}$ in ^{43}Ca$^+$ [19].

3.3.1 Intermediate Field

In the above calculations we have assumed that F and m_F are good quantum numbers. However this assumption starts to break down at 146 G and we can no longer treat the Zeeman interaction as a mere perturbation but instead need to find the eigenvalues of the total Hamiltonian

$$H = AI \cdot J - \mu_J \cdot B - \mu_I \cdot B \qquad (3.37)$$

The total electronic magnetic moment is $\mu_J = -g_J\mu_B J$ and the nuclear magnetic moment is $\mu_I = g_I\mu_N I$. The hyperfine coupling constant A has to be measured empirically. The mixing of the levels leads to a correction of the matrix elements given in Table 3.1. While this correction is considerable in the D states, the hyperfine structure in the S and P states is large enough compared to the Zeeman splitting that the correction is only a few percent.

For $|F = 4, m_F = 0\rangle \leftrightarrow |F = 3, m_F = 1\rangle$ the intermediate field corrections cause the transition to become a first-order B insensitive clock transition. Therefore, as for all clock transitions, $\Gamma_{el} = 0$ and to a good approximation $\Omega_{Raman,\downarrow} = \Omega_{Raman,\uparrow}$ [18, 20].

3.4 Geometric Phase Gate

The most complex gate in the universal gate set introduced in Chap. 1 is the CNOT gate that can entangle two qubits with each other. With trapped ions we typically implement a controlled phase gate $\begin{pmatrix} 1 & & & \\ & i & & \\ & & i & \\ & & & 1 \end{pmatrix}$, that can easily be transformed into a CNOT gate using single-qubit operations. In order to create entanglement between two particles they need to be able to interact with each other—either directly or via another entangled particle. While the electronic wave-functions of two ions in the same trap are well separated, ions are coupled strongly via their motion due to their equal repulsive charge. We can excite an ion's motion using laser fields, and we can tailor the laser fields such that motional excitation depends on the ion's spin state. This conditional excitation and the motional coupling between the ions can be used for implementing an entangling gate. There exist several schemes for entangling gates with trapped ions that are based on this principle [21–25]. We implement the σ_z geometric phase gate first proposed and implemented in [23], and roughly follow the theoretical derivations in [5, 14, 20, 23, 25].

3.4.1 Creating a Spin-Dependent Force

For the stretch qubit, a pair of Raman beams with polarisations $\sigma^\pm$, $\sigma^\mp$ will not couple two levels of opposite spin to each other, but will instead couple to each level independently via light shifts. When choosing a frequency difference of $\nu = \delta_r - \delta_b \approx \omega_z$ this will lead to a near-resonant force exciting the ion's motion. Assuming that the difference vector of the two Raman lasers Δk is parallel to the trap axis $\hat{z}$ and therefore only couples to the axial motion, the interaction Hamiltonian for a single ion in state $|1\rangle$ will look like

$$H_I(t) = -\frac{\hbar\Omega_{R,r}}{2}e^{i(\delta_r t - k_r \cdot z + \phi_r)} |3\rangle\langle 1| - \frac{\hbar\Omega_{R,b}}{2}e^{i(\delta_b t - k_b \cdot z + \phi_b)} |3\rangle\langle 1| + \text{h.c.} \quad (3.38)$$

and after applying the James-Jerke approximation analogously to Sect. 3.2.3, neglecting constant light-shift terms and shifts on the excited level $|3\rangle$ only, we obtain

$$H_{\text{eff}}(t) = -\frac{\hbar \Omega_r \Omega_b}{4 \Delta_R} |1\rangle\langle 1| e^{i(\Delta k.z - \nu t - \Delta\phi)} + \text{h.c.} \qquad (3.39)$$

where $\nu = \delta_r - \delta_b$, and $\Omega_s = \frac{\Omega_r \Omega_b}{2\Delta}$ depends on the state $|s\rangle = |1\rangle$ the laser is coupling to. For an entangling gate the ion is prepared in an equal superposition of the two qubit states $|\uparrow\rangle$ and $|\downarrow\rangle$, and we need to sum over the coupling to both of those states. We will further write the initial phase difference of the two Raman beams as $\Delta\phi = -\phi_0$:

$$H_{\text{eff}}(t) = \sum_{s=\uparrow,\downarrow} -\frac{\hbar \Omega_s}{2} |s\rangle\langle s| e^{i(\Delta k.z - \nu t + \phi_0)} + \text{h.c.} \qquad (3.40)$$

With $\Delta k.z = \eta(a + a^\dagger)$ and expanding the exponential to first order in the Lamb-Dicke approximation we obtain

$$H_{\text{eff}}(t) = \sum_{s=\uparrow,\downarrow} -\frac{\hbar \Omega_s}{2} |s\rangle\langle s| \left[2\cos(\nu t - \phi_0) + 2\eta(a + a^\dagger)\sin(\nu t - \phi_0) \right] \qquad (3.41)$$

The first term corresponds to a light-shift that leads to a single-qubit phase whereas the second term has the shape of a harmonic driving force, which will cause a displacement in phase-space [26]. For most gates where $\Omega_s \ll \nu$ we can neglect the time-dependent light shift, as the acquired single-qubit phase will then be small, and errors caused by it can be suppressed with pulse-shaping [5].

After moving into the interaction picture of the harmonic oscillator with $a \mapsto ae^{-i\omega_z t}$, $a^\dagger \mapsto a^\dagger e^{i\omega_z t}$ we obtain

$$\hat{H}_{\text{eff}}(t) = \sum_{s=\uparrow,\downarrow} -\frac{\hbar \Omega_s}{2} |s\rangle\langle s| \left[i\eta \left\{ ae^{-i[(\omega_z + \nu)t - \phi_0]} - a^\dagger e^{i[(\omega_z + \nu)t - \phi_0]} \right. \right.$$
$$\left. \left. - ae^{-i[(\omega_z - \nu)t + \phi_0]} + a^\dagger e^{i[(\omega_z - \nu)t + \phi_0]} \right\} \right] \qquad (3.42)$$

For $\nu = \omega_z + \delta_g$, $\delta_g \ll \omega_z$ we can neglect the terms that are fast rotating with $\omega_z + \nu$ in the rotating wave approximation. The Hamiltonian then simplifies to

$$\hat{H}_{\text{eff}}(t) = \sum_{s=\uparrow,\downarrow} \frac{\hbar \Omega_s}{2} i\eta |s\rangle\langle s| \left[ae^{i(\delta_g t - \phi_0)} - a^\dagger e^{-i(\delta_g t - \phi_0)} \right] \qquad (3.43)$$

This Hamiltonian gives rise to a force that acts on each spin-state with a different state-dependent force amplitude Ω_s. The dynamics become more evident when we calculate the propagator $U(t)$ using the Magnus expansion (see Appendix C.1). We obtain

$$U(t) = \left(\sum_{s=\uparrow,\downarrow} D\left(\alpha_s\right) |s \rangle\langle s| \right) e^{-i\Phi} \tag{3.44}$$

$$\alpha_s = -\frac{\Omega_s \eta}{\delta_g} e^{-i\phi_0} e^{-i\delta_g t/2} \sin\left(\frac{\delta_g t}{2}\right)$$

$$\Phi = \frac{\eta^2}{4\delta_g^2} [\delta_g t - \sin(\delta_g t)] \left(\sum_{s=\uparrow,\downarrow} \Omega_s |s \rangle\langle s| \right)^2$$

There are two effects on the ion: (i) a displacement of the state $|s\rangle$ by α_s and (ii) accumulation of the phase Φ. The displacement of the spin-state $|s\rangle$ is proportional to Ω_s and we can see from Table 3.1 that Ω_s has a different sign for opposite spin states $|\uparrow\rangle$ and $|\downarrow\rangle$, i.e. the force acting on the different spin states is opposite. This conditionality of the force on the ion's spin gives rise to the non-linearity that is a key requirement for the creation of entanglement. For $\delta_g = 0$ the force is resonant with the motion and if applied to an ion in a superposition of $|\uparrow\rangle$ and $|\downarrow\rangle$ it drives the two components further apart from each other in motional phase-space, creating a large 'cat-state'. For $\delta_g \neq 0$ the displacement is periodic with $\sin\left(\frac{\delta_g t}{2}\right)$ and the ion's motion is first excited and then attenuated again. As the displacement depends on $|s\rangle$, $\alpha_s \neq 0$ means that the spin and motion of a single ion are entangled. This is in general not desirable, because then finite ion temperature and heating would lead to decoherence of the internal state. For an entangling gate we therefore want $\frac{\delta_g t_g}{2} = k\pi$, with k a natural number, so that at the end of the gate ($t = t_g$) $\alpha_s = 0$. This means the only effect of the propagator $U(t)$ is accumulation of the phase Φ. For $\Omega_\uparrow = \Omega_\downarrow$ the operator $\sum_s \Omega_s |s\rangle\langle s|$ is identity and the phase Φ is a simple global phase. For $\Omega_\uparrow = -\Omega_\downarrow$ however we obtain a σ_z operator, leading to a relative phase difference between the two different spin states. For a single ion this phase is of no special interest and can easily be mimicked by other, simpler operations. For two ions however $\Phi = \pi$ corresponds to a conditional phase gate, that can be used to create a maximally entangled Bell state.

For clock qubits $\Omega_{\text{Raman},\downarrow} = \Omega_{\text{Raman},\uparrow}$. This means the force is not spin-dependent and can therefore not be used to form an entangling gate, but only leads to acquisition of a global phase.[5] A related gate mechanism however, the Mølmer-Sørensen gate [22], works well on clock qubits. A parallel derivation and comparison of the σ_z phase gate and the Mølmer-Sørensen gate can be found in [20, 25].

The Geometric Phase

It is interesting to note the geometric nature of the phase Φ. In first order the Magnus expansion only leads to the displacement operator $U(t) = \sum_s D(\alpha_s) |s\rangle\langle s|$. The phase arises because the Hamiltonian at different times does not commute

---

[5]In higher order terms there is a deviation in the Rabi frequencies on the two spin-states [20]. However this difference is so small, that an entangling gate would be highly inefficient and is therefore not feasible.

with itself. With the identity $D(\alpha)D(\beta) = D(\alpha + \beta)e^{i\text{Im}(\alpha\beta^*)}$ and concatenating many infinitesimal displacements one obtains the above propagator $U(t) = D(\alpha)e^{i\Phi}$ where for γ a closed path through phase-space [23]

$$\Phi = \text{Im}\left(\int_{\gamma} \alpha^* d\alpha\right) = \frac{1}{\hbar}\int_A dz dp = \frac{A}{\hbar} \tag{3.45}$$

This means the phase Φ only depends on the area A enclosed in phase space, but not on the time necessary to traverse this trajectory or on the speed.[6]

3.4.2 Two Ion Geometric Phase Gate

In applying this Hamiltonian to two ions we have to consider several things: (i) The ions sit at different positions z in the standing wave and therefore experience different phases $\phi_z = k.z_0$ of the laser light. So far we have absorbed this phase into $\phi_{r/b} \stackrel{\frown}{=} \phi_0$, now however we have to distinguish between (a) ϕ_0, the relative phase difference of the two Raman beams at the beginning of the gate. It is identical for both ions but changes over time between different experiments. (b) ϕ_z, the phase difference between the two ions due to their different positions which does not change over time. We can adjust ϕ_z via the ion spacing. (ii) For two ions there are two axial normal modes, the center-of-mass or in-phase (ip) mode with $z_1 = z_2$ and the stretch, breathing, or out-of-phase mode (oop) with $z_1 = -z_2$. The corresponding mode vector $\xi_m = (\xi_{m,1}, \xi_{m,2})$ for the two ion crystal is $\xi_{ip} = \frac{1}{\sqrt{2}}(1, 1)$ for the ip-mode and $\xi_{oop} = \frac{1}{\sqrt{2}}(1, -1)$ for the oop-mode. Depending on the laser detuning ν, the ion spacing and the ion's spin state we can excite either mode. We label the ladder operators for the different modes $a_m^{\dagger}$ with $m = $ ip, oop. The Lamb-Dicke parameter of the two modes differ due to the different mode frequencies $\eta_{oop} = \eta_{ip}/3^{1/4}$. (iii) The force exerted by the light field on the ions is the sum of the force on each ion. It is therefore twice as large as for a single ion if the forces are in phase. However the Lamb-Dicke effect for two ions is smaller $\eta_{2ions} = \eta_{1ion}/\sqrt{2}$, cancelling part of the effect.

The Hamiltonian for two ions equivalent to Eq. (3.42) then looks like[7]

$$\hat{H}(t) = \sum_{m=ip,oop} \sum_{j=1,2} \sum_{s_j=\uparrow,\downarrow} -\frac{\hbar}{2}\left(\xi_{m,j}e^{i\phi_{z,j}}\Omega_{s_j}\right)|s_j\rangle\langle s_j|_j \cdot$$

$$\left[i\eta_m\left\{a_m e^{-i[(\omega_{z,m}+\nu)t-\phi_0]} - a_m^{\dagger}e^{i[(\omega_{z,m}+\nu)t-\phi_0]}\right.\right.$$

$$\left.\left. -a_m e^{-i[(\omega_{z,m}-\nu)t+\phi_0]} + a_m^{\dagger}e^{i[(\omega_{z,m}-\nu)t+\phi_0]}\right\}\right] \tag{3.46}$$

---

[6]More accurately, the two-qubit phase also has a dynamic phase component. However the dynamic phase is exactly twice the geometric phase with opposite sign, resulting in a final two-qubit phase corresponding to the geometric phase, see [27].

[7]We have assumed $\phi_{z,j} = 0, \pi, \dots$ for factoring it out. This is the case in all our experiments.

Here $j = 1, 2$ is the index of the respective ions, and we set the global phase such that the phase $(\xi_{m,1} e^{i\phi_{z,1}}) = 1$ for ion 1 with $\phi_{z,1} = 0$ and $\phi_{z,2} = \phi_z$. The spin operator $|s_j \rangle\langle s_j|_j$ is $|s_1 \rangle\langle s_1| \otimes I$ for $j = 1$ and $I \otimes |s_2 \rangle\langle s_2|$ for $j = 2$. For $\delta_g \ll \omega_z$ we can neglect excitation of the second mode and our final Hamiltonian after the rotating wave approximation for a gate on the in-phase mode becomes

$$\hat{H}(t) = \sum_{s_1,s_2=\uparrow,\downarrow} \frac{i\eta\hbar}{2} \left(\Omega_{s_1} + e^{i\phi_z}\Omega_{s_2}\right) |s_1s_2 \rangle\langle s_1s_2| \left[ae^{i(\delta_g t - \phi_0)} - a^\dagger e^{-i(\delta_g t - \phi_0)}\right]$$

(3.47)

For each δ_g with $\frac{\delta_g t_g}{2} = \pi$ we can find a laser power P such that

$$\frac{\eta^2 t_g}{4\delta_g} \left[\left(\Omega_s + e^{i\phi_z}\Omega_{-s}\right)^2 - \left(\Omega_s + e^{i\phi_z}\Omega_s\right)^2\right] = \pi/2 \qquad (3.48)$$

The ion spacing in this thesis was set to be a half-integer multiple of the standing wave periodicity, i.e. $\delta z = (k + \frac{1}{2})\lambda_z$, and therefore $e^{i\phi_z} = -1$. For $\Omega_\uparrow = -\Omega_\downarrow$ and an initial spin state $|\uparrow\uparrow\rangle$ or $|\downarrow\downarrow\rangle$ the forces cancel and we have no displacement in phase space and therefore no phase accumulation. For initial spin states $|\uparrow\downarrow\rangle$ or $|\downarrow\uparrow\rangle$ however the force amplitudes are $\pm 2\Omega_\uparrow$, and both states acquire an equal geometric phase Φ_s. For $\Phi_s = \pi/2$ this leads to the propagator

$$U(t_g) = \begin{pmatrix} 1 & & & \\ & i & & \\ & & i & \\ & & & 1 \end{pmatrix} \qquad (3.49)$$

This operator corresponds to the desired CPHASE propagator. Typically this gate is performed within a Ramsey interferometer, which leads to the creation of a Bell-state via

$$R\left(\frac{\pi}{2}\right) U_{\text{gate}} R\left(\frac{\pi}{2}\right) |\downarrow\downarrow\rangle = \frac{1}{\sqrt{2}} \left(|\downarrow\downarrow\rangle + i|\uparrow\uparrow\rangle\right) \qquad (3.50)$$

with rotation matrix $R(\phi) = R_x(\phi) \otimes R_x(\phi)$, and $R_x(\phi)$ as defined in Appendix A. The total gate operation is illustrated in Fig. 3.4.

Asymmetric Rabi frequencies. For $|\Omega_\uparrow| \neq |\Omega_\downarrow|$, as is the case in $^{43}\text{Ca}^+$, an additional single qubit phase is accumulated during the gate operation, see Appendix C.2. This phase can be cancelled by applying a π-pulse in the center of the gate and therefore distributing the gate on two arms of a spin echo sequence. Apart from reducing errors due to magnetic field noise, this also flips all spins mid-way and therefore all asymmetries (e.g. due to unequal illumination of the ions) in the geometric phase accumulation are cancelled. When distributing the gate over two arms

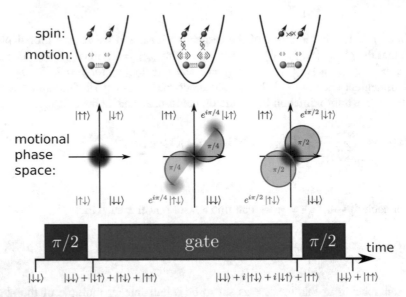

Fig. 3.4 The geometric phase gate: The ions are prepared in $|\downarrow\downarrow\rangle$ and then transferred to $\frac{1}{2}(|\downarrow\downarrow\rangle + |\downarrow\uparrow\rangle + |\uparrow\downarrow\rangle + |\uparrow\uparrow\rangle)$ via a $\pi/2$-pulse. At this time all spin and motion degrees are disentangled, the ions are cooled close to their ground state and are located at the origin in motional phase space. Application of the gate-pulse excites the motion of the ions and leads to entanglement of the spin and motional degrees of each ion. While the spin-components $|\downarrow\downarrow\rangle$ and $|\uparrow\uparrow\rangle$ (black) are not displaced and remain at the origin, the components $|\downarrow\uparrow\rangle$ and $|\uparrow\downarrow\rangle$ (blue/green) experience a force that leads to a circular displacement in phase space in the rotating frame. In the process they accumulate a geometric phase Φ_s, corresponding to the signed area enclosed by their trajectory. After the gate time t_g the motion of the ions has been fully attenuated again and all spin components have returned to the origin. The spin and motion of each ion are disentangled again, however the accumulated geometric phase $e^{i\pi/2} = i$ has led to entanglement of the spin degrees of the two ions. A second $\pi/2$-pulse finally projects the ions' state onto a Bell state

of a spin-echo sequence, each arm has to contain a full loop in phase space, i.e. in each arm $t_g = \frac{2\pi}{\delta_g}$ and the total gate time is $2t_g$. The power is reduced by $\sqrt{2}$ such that the total geometric phase accumulated in both arms together is $\Phi = \pi$. To ensure the force is continuous over the two arms, the phase of the second part of the gate has to be adjusted to match the delay due to the first part, the π-pulse, any padding delays and the phase-flip due to the π-pulse to $\varphi = \varphi_0 + \delta_g \cdot (t_g + t_\pi + t_{\text{padding}}) - \pi$.

Choosing the Raman polarisation. The magnitude and sign of $\Omega_\uparrow$, $\Omega_\downarrow$ depend on the polarisation of the Raman beams. For a geometric phase gate we need $\Omega_\uparrow \neq \Omega_\downarrow$ and for optimal power-efficiency we want maximal $|\Omega_\uparrow - \Omega_\downarrow|$ and $\text{sign}(\Omega_\uparrow) = -\text{sign}(\Omega_\downarrow)$. For arbitrary polarisations we have $\Omega_\uparrow - \Omega_\downarrow = \frac{7\omega_f(b_+ r_+ - b_- r_-)}{24\Delta(\Delta - \omega_f)}$ for ^{43}Ca$^+$ and $\Omega_\uparrow - \Omega_\downarrow = \frac{\omega_f(b_- r_- - b_+ r_+)}{3\Delta(\Delta - \omega_f)}$ for ^{88}Sr$^+$. We further want no differential light shift from both Raman beams, i.e. $e_+^2 - e_-^2 = 0$. The ideal polarisations are therefore $\hat{\sigma}^\pm$ for one and $\hat{\sigma}^\mp$ for the other Raman beam for both ion species.[8] These polarisations give the light shift forces listed in Table 3.1.

3.5 Mixed Species Gate

A useful characteristic of the σ_z geometric phase gate is that it is independent of the qubit frequency ω_0. That means we can in principle use the same driving force for different species of ion. However there are several complicating factors:

(i) **Effects of wavelength**: The wavelengths for different atomic species are usually very different. A laser hence typically only couples to one atomic species, and will be far off-resonant for any other species.

(ii) **Effects of mass**: Different atomic species possess different mass. The larger the difference of the masses, the more asymmetric is the motional excitation of the two ions. The different masses also lead to different Lamb-Dicke factors for the two ions.

(iii) **Effects of level structure**: If we can find a laser that couples equally to both species, the matrix elements for different species also vary if the atomic level structures differ. This leads to asymmetries in the Rabi-frequencies for the two ions.

By thoughtful choice of the ions' species, adaptation of the gate sequence and control of additional parameters we can mitigate many of these effects and achieve high fidelity entanglement of different species of ion.

Common Raman beams Although we could in principle use two different pairs of Raman beams to create the spin-dependent force and avoid the problem described in (i), we note that the $\lambda = 408$ nm transition between $5S_{1/2} \leftrightarrow 5P_{3/2}$ in ^{88}Sr$^+$ is only $\Delta = 2\pi \cdot 20$ THz detuned from the $\lambda = 397$ nm transition between $4S_{1/2} \leftrightarrow 4P_{1/2}$ in ^{43}Ca$^+$, see Fig. 3.5. To reduce photon scattering errors to the 10^{-4} level a Raman detuning of $\Delta \gtrsim 2\pi \cdot 10$ THz is required. Using a single pair of Raman beams tuned between the $\lambda = 408$ nm transition in ^{88}Sr$^+$ and the $\lambda = 397$ nm in ^{43}Ca$^+$ is therefore ideal in terms of scattering, and also greatly reduces experimental complexity.

Motional coupling The motion and sympathetic cooling of mixed-species crystals is covered in detail in [9, 28]. For effective cooling and motional coupling it is desirable to choose species with masses as similar as possible. The masses of ^{43}Ca$^+$ and

---

[8]When including the hyperfine splitting into the Raman detuning the matrix elements don't cancel exactly any more. This means for ^{43}Ca$^+$ a small asymmetry in the $\hat{\sigma}^+$ and $\hat{\sigma}^-$ polarisation components would be necessary to null the single beam light shift. However this would then cause a differential light shift for ^{88}Sr$^+$ which does not have a hyperfine structure. In practice this light shift is small enough that it doesn't cause a significant error and the polarisation is set to null the light shift for ^{43}Ca$^+$.

Fig. 3.5 Mixed species gate:
A single laser at $\lambda \approx 402$ nm
detuned by
$\Delta_{Ca} \approx \Delta_{Sr} \approx 2\pi \cdot 10$ THz
from transitions in both
^{43}Ca$^+$ and ^{88}Sr$^+$ can be used
to drive a geometric phase
gate on both species
simultaneously. The
fine-structure splittings are
$\omega_{f,Ca} = 2\pi \cdot 6.68$ THz and
$\omega_{f,Sr} = 2\pi \cdot 24.03$ THz

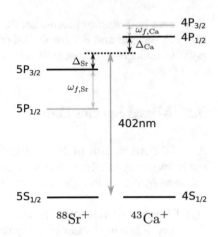

^{88}Sr$^+$ only differ by a factor of 2 and are therefore relatively well suited. Gates should be performed on the axial rather than the radial modes, because axial mode frequencies for different species are more similar and therefore the mode excitation amplitudes for the two ions are more balanced.

While for same-species crystals the oop-mode is purely anti-symmetric and therefore is not heated by noisy uniform electric fields, this is no longer true for mixed species crystals. Here only crystals with odd numbers of ions can exhibit purely anti-symmetric modes of motion, and the motional heating-rate of the oop-mode is therefore larger in mixed-species compared to same-species crystals.

The Lamb-Dicke factor in axial direction in our beam geometry is now [28]

$$\tilde{\eta}_{j,m} = \sqrt{\frac{\hbar}{M_j \omega_m}} k \xi_{j,m} \qquad (3.51)$$

where $j = 1, 2$ is the ion's index, $m =$ ip, oop the motional mode and factors due to the Raman beam geometry are already included so that $k = 2\pi/\lambda$. For two ions the mode eigenvectors $\xi_{j,m}$ can be calculated analytically [9]. The values for a ^{88}Sr$^+ - {}^{43}$Ca$^+$ crystal are tabulated in Table 3.2. We typically extract the mode vector from the Lamb-Dicke parameter and write $\tilde{\eta}_{j,m} = \eta_{j,m} \xi_{j,m}$.

Table 3.2 Axial mode eigenvectors of a ^{43}Ca$^+ - {}^{88}$Sr$^+$ crystal: The eigenvectors of the axial modes depend only on the ratio of the masses of the two species. For a same-species crystal they are $(\xi_{z1,ip}, \xi_{z2,ip}) = \frac{1}{\sqrt{2}}(1, 1)$ and $(\xi_{z1,oop}, \xi_{z2,oop}) = \frac{1}{\sqrt{2}}(-1, 1)$

Mode	^{43}Ca$^+$ $\xi_{z1,m}$	^{88}Sr$^+$ $\xi_{z2,m}$
ip	0.453	0.892
oop	−0.892	0.453

Ion order Electric or magnetic field gradients and inhomogeneous stray fields can cause changes in the qubit frequency depending on ion position [28]. They can also lead to slight displacement of the ions, which is larger for heavier ions, and gives rise to changes in the motional frequencies and micromotion. It is therefore important to keep the order of ion species in the crystal fixed.

Sympathetic cooling While the lowest attainable temperature for sympathetically cooled mixed-species crystals is in principle the same as for same-species crystals, the cooling efficiency will be lower due to asymmetry in the mode amplitudes and cooling is therefore slower. Additionally the large mode amplitude differences for the radial modes mean that radial modes of the sympathetically cooled ion species will be cooled inefficiently and may therefore stay fairly hot.

Rabi frequency asymmetry For mixed species $\eta \xi_j \Omega_{s_j}$ in Eq. (3.46) has to be changed to $\eta_j \xi_j \Omega_{s_j}$, where now the magnitude of both η and ξ differ for the two ions. This and differences in $\Delta_{^{43}Ca^+}$ and $\Delta_{^{88}Sr^+}$ lead to a strong asymmetry in Rabi frequencies. To symmetrise the gate function and ensure that both odd (even) spin configurations accumulate the same geometric phase, we implement the gate in a two-loop sequence divided over both arms of a spin-echo sequence as described in Sect. 3.4.2. The total accumulated phase is then a trivial global phase accumulated equally on all initial spin-states plus the geometric two-qubit phase we are interested in. The two-qubit phase now corresponds to the difference of the areas enclosed in phase space of even and odd spin states, and for motional mode m we obtain:

$$\Phi \propto (\Phi_{\text{even}} - \Phi_{\text{odd}}) \tag{3.52}$$

$$= \left[\left(\eta_{1,m} \xi_{1,m} \Omega_{1,\uparrow} + e^{i\phi_z} \eta_{2,m} \xi_{2,m} \Omega_{2,\uparrow} \right)^2 \right.$$
$$\left. + \left(\eta_{1,m} \xi_{1,m} \Omega_{1,\downarrow} + e^{i\phi_z} \eta_{2,m} \xi_{2,m} \Omega_{2,\downarrow} \right)^2 \right]$$
$$- \left[\left(\eta_{1,m} \xi_{1,m} \Omega_{1,\uparrow} + e^{i\phi_z} \eta_{2,m} \xi_{2,m} \Omega_{2,\downarrow} \right)^2 \right.$$
$$\left. + \left(\eta_{1,m} \xi_{1,m} \Omega_{1,\downarrow} + e^{i\phi_z} \eta_{2,m} \xi_{2,m} \Omega_{2,\uparrow} \right)^2 \right]$$

Accumulation of the additional global phase means that the gate mechanism is driven less efficiently and therefore larger Raman laser intensity is required. Because the loops in phase-space are larger this also means the gate becomes more sensitive to certain sources or error, for example heating or out-of-Lamb-Dicke effects. This is discussed in detail in Sect. 3.7. For $\Omega_{\uparrow,1} = \alpha \Omega_{\uparrow,2}$ the gate efficiency $\zeta = (\Phi_{\text{odd}} - \Phi_{\text{even}}) / (\Phi_{\text{odd}} + \Phi_{\text{even}})$ scales as

$$\zeta = \frac{2\alpha}{1 + \alpha^2} \tag{3.53}$$

3.6 Fast Geometric Phase Gate

We can reduce the gate duration $t_g = \frac{2\pi}{\delta_g}$ by increasing the gate detuning δ_g. However, once we get close to $t_g \approx \frac{1}{\omega_z}$ problems arise, because several assumptions we made start to fail:

(i) The gate detuning becomes so large that excitation of the second axial mode is no longer negligible. In general the loops in phase space of different motional modes do not close at the same time.
The remaining spin-motion entanglement and displacement in phase-space lead to significant errors.

(ii) The rotating wave approximation starts to fail and counter-rotating terms can no longer be neglected. Due to the breakdown of the rotating wave approximation the Raman phase ϕ_0 no longer only affects the orientation of the trajectories in phase-space, but also their shape and hence the magnitude of the geometric phase.

(iii) The time-dependent light-shift in Eq. (3.41) becomes too large to ignore. This light shift also depends on ϕ_0.

In recent years several methods have been proposed to overcome these problems [29–33]. We follow the proposal by Steane et al. [32] that uses cw-pulses shaped in amplitude to perform gates at timescales down to and below a single motional period of the ions. The shaping of the amplitude introduces new degrees of freedom that allow us to choose sequences that are robust to, or strongly suppress, the above-mentioned error sources. Following Steane's proposal, sequences are chosen that minimise the coherent error of the gate while fulfilling the boundary conditions that

(i) at the end of the sequence the displacement in phase-space $\Delta\alpha = 0 \; \forall\phi_0$,
(ii) the acquired geometric phase Φ is independent of the initial phase of the driving field ϕ_0 and
(iii) the single-qubit phase θ_{LS} due to the AC Stark shifts vanishes $\forall\phi_0$.

3.6.1 Gate Dynamics

We follow the theoretical derivation in [32], while additionally allowing phase-chirps in the laser control field. We begin with Hamiltonian (3.41) for two ions, where we have made the Lamb-Dicke approximation, but no rotating wave approximation:

$$H_I(t) = \sum_{j=1,2} \sum_{s_j=\uparrow,\downarrow} -\hbar \left(e^{i\phi_{z,j}} \Omega_{s_j} \right) |s_j\rangle\langle s_j|_j \cdot$$

$$\left[\cos(\nu t - \phi_0) + \sum_{m=\text{ip,oop}} \xi_{m,j}\eta_m (a_m + a_m^\dagger) \sin(\nu t - \phi_0) \right]$$

(3.54)

While we can not neglect the first (light shift) part of the Hamiltonian, it does commute with the second part of the Hamiltonian and we can therefore investigate the dynamics of the two parts independently.

Part I: Driven Harmonic Oscillator

As shown in Carruthers and Nieto [26] the second part of the Hamiltonian, which is of the form $-z_0(a + a^\dagger)F(t)$, corresponds to a driven harmonic oscillator with position-independent driving force

$$F_{m,|s_1 s_2\rangle}(t) = 4M\omega_m z_{0,m} \Omega_{m,|s_1 s_2\rangle}(t) \sin(\nu t - \phi_0 + \varphi_{\text{ch}}(t))$$ (3.55)

where we have introduced $\Omega_{m,|s_1 s_2\rangle}(t) = \frac{\eta_m}{2} \left[\Omega_{s_1}(t) + \xi_{m,2} e^{i\phi_z} \Omega_{s_2}(t) \right]$, ground-state wave function spread $z_{0,m} = \sqrt{\frac{\hbar}{2M\omega_m}}$ and allowed for a phase-chirp $\varphi_{\text{ch}}(t)$. We can rewrite the phase chirp as $\Omega \sin(\nu t + \varphi_{\text{ch}}(t)) = \Omega_i \sin(\nu t) + \Omega_q(t) \cos(\nu t)$, where the quadrature amplitude $\Omega_q(t)$ describes the magnitude of the phase chirp at a certain time. As shown in [26], we can estimate the displacement in motional phase space caused by this force with

$$\Delta\alpha_{m,|s_1 s_2\rangle}(t) = \frac{i}{\sqrt{2M\omega_m \hbar}} \int_0^t e^{i\omega_m t'} F_{m,|s_1 s_2\rangle}(t') dt' = e^{i\phi_0} \Delta\alpha_{m,|s_1 s_2\rangle}^+ + e^{-i\phi_0} \Delta\alpha_{m,|s_1 s_2\rangle}^-$$

$$\Delta\alpha_{m,|s_1 s_2\rangle}^+ = \int_0^t e^{i\delta_+ t'} \left(\Omega_{i,m,|s_1 s_2\rangle}(t') + i\Omega_{q,m,|s_1 s_2\rangle}(t') \right) dt'$$

$$\Delta\alpha_{m,|s_1 s_2\rangle}^- = -\int_0^t e^{i\delta_- t'} \left(\Omega_{i,m,|s_1 s_2\rangle}(t') - i\Omega_{q,m,|s_1 s_2\rangle}(t') \right) dt'$$ (3.56)

where $\delta_{\pm,m} = \omega_m \pm \nu$. Analogous to the gate in the slow, adiabatic regime we can calculate the geometric phase acquired due to the displacement in phase space by

$$\Phi = \sum_m \sum_s \text{Im}\left(\int_{\gamma_{m,s}} \alpha_{m,s}^* d\alpha_{m,s} \right) = \sum_m \sum_s \text{Im}\left(\int_0^t \alpha_{m,s}^* \dot{\alpha}_{m,s}(t') dt' \right) = \sum_m \sum_s \text{Im}(I_{m,s})$$

(3.57)

This geometric phase Φ has a two-qubit phase component Ψ—which is the source of entanglement creation—as well as a single-qubit phase component Θ if there is an asymmetry in the magnitude of the Rabi frequencies Ω_s. Definitions for the single-qubit and two-qubit phases can be found in Appendix C.4. Because each spin state

$s = |s_1 s_2\rangle$ experiences a displacement in both the ip and oop mode, we have to sum over both modes to acquire the total phase. Depending on the handedness in which the path is traversed, which depends for example on the Raman beat note frequency ν relative to the mode frequency ω_m, these phases can add or subtract, affecting the efficiency of the gate. Here

$$I_{m,s} = e^{2i\phi_0} I_{m,s}^+ + e^{-2i\phi_0} I_{m,s}^- + I_{m,s}^0 \tag{3.58}$$

$$I_{m,s}^0 = \int_0^t \Delta\alpha_{m,s}^{+\,*} e^{i\delta_{+,m}t'} \left[\Omega_{i,m,s}(t') + i\Omega_{q,m,s}(t')\right] - \Delta\alpha_{m,s}^{-\,*} e^{i\delta_{-,m}t'} \left[\Omega_{i,m,s}(t') - i\Omega_{q,m,s}(t')\right] \mathrm{d}t'$$

$$I_{m,s}^- = -\int_0^t \Delta\alpha_{m,s}^{+\,*} e^{i\delta_{-,m}t'} \left[\Omega_{i,m,s}(t') - i\Omega_{q,m,s}(t')\right] \mathrm{d}t'$$

$$I_{m,s}^+ = \int_0^t \Delta\alpha_{m,s}^{-\,*} e^{i\delta_{+,m}t'} \left[\Omega_{i,m,s}(t') + i\Omega_{q,m,s}(t')\right] \mathrm{d}t'$$

With these equations we can calculate the motional displacement and accumulated geometric phase for two ions due to a σ_z spin-dependent force in the non-adiabatic regime.

3.6.1.1 Part II: Time-Dependent Light Shift

We now look at the remaining first part of the Hamiltonian, that corresponds to a time-dependent light-shift. For $e^{i\phi_z} = -1$ the Hamiltonian simplifies to

$$H_{I,\mathrm{LS}}(t) = -\hbar \left[\frac{\Omega_\uparrow(t) - \Omega_\downarrow(t)}{2}(\sigma_z \otimes I) + \frac{\Omega_\downarrow(t) - \Omega_\uparrow(t)}{2}(I \otimes \sigma_z) \right] \cos(\nu t - \phi_0) \tag{3.59}$$

leading to the propagator

$$U_{\mathrm{LS}}(t) = e^{i\theta_{\mathrm{LS}}\sigma_{z,1}/2} e^{-i\theta_{\mathrm{LS}}\sigma_{z,2}/2} \tag{3.60}$$

$$\theta_{\mathrm{LS}} = e^{-i\phi_0}\vartheta^+ + e^{i\phi_0}\vartheta^{+*}$$

$$\vartheta^+ = \int_0^{t_g} \frac{\Omega_\uparrow(t) - \Omega_\downarrow(t)}{2} e^{i\nu t} \mathrm{d}t$$

with Rabi frequency $\Omega_s(t) = \Omega_{i,s}(t) + i\Omega_{q,s}(t)$, see also [5]. Because ϕ_0 is not controlled in the experiment, and varies from run to run, this single-qubit phase takes random values. The Rabi frequency in θ_{LS} is not suppressed by η as it is for the geometric phase Φ. The magnitude of θ_{LS} can therefore be large which would lead to considerable errors.

From these quantities we can numerically integrate the entire dynamics of the gate. If we choose the Rabi frequency such that the gate sequence consists of a concatenation of rectangular top-hat functions, we can replace the integrals with sums and obtain analytic expressions for all quantities, see [32].

3.6.2 Estimating Errors

Still following [32], we can get a rough estimate of the gate error by considering a generalised gate operation $U_{\text{gate}} = D(\alpha)U_{\sigma_z \otimes \sigma_z}(\psi/2)U_{\sigma_z \otimes I}(\theta_1/2)U_{I \otimes \sigma_z}(\theta_2/2)$, where for an ideal gate $\alpha = \theta_1 = \theta_2 = 0$ and $\psi = \pi/2$. An imperfect gate operation leads to a state

$$U_{\text{exp}} |s_1 s_2\rangle |0, n_{\text{ip}}; 0, n_{\text{oop}}\rangle = e^{i(\theta_1/2\sigma_{z,1} + \theta_2/2\sigma_{z,2} + \psi/2\sigma_{z,z})} |s_1 s_2\rangle |\Delta\alpha_{\text{ip}}, n_{\text{ip}}; \Delta\alpha_{\text{oop}}, n_{\text{oop}}\rangle$$

$$= \sum_i e^{i\Phi_{si}} |s_1 s_2\rangle_i |\Delta\alpha_{\text{ip}_i}, n_{\text{ip}}; \Delta\alpha_{\text{oop}_i}, n_{\text{oop}}\rangle \qquad (3.61)$$

To first order, and therefore ignoring all mixed terms, we can estimate the gate error as

$$\epsilon = 1 - \left\langle \left| \langle\varphi| U_{\text{exp}}^\dagger U_{\text{id}} |\varphi\rangle \right|^2 \right\rangle_{\text{th}} \qquad (3.62)$$

$$\approx \sum_{m,s} \frac{1 + 2\bar{n}_m}{4} |\Delta\alpha_m^s|^2 + \frac{\Delta\Psi^2}{8} + \frac{\Delta\theta_1^2 + \Delta\theta_2^2}{4}$$

for an input state $|\phi\rangle = \frac{1}{2}(1, 1, 1, 1)$, with thermal average over the motional states $\langle\rangle_{\text{th}}$, see Appendix C.3. The input state $|\phi\rangle$ is the one used in our experiments and is propagated to a maximally entangled state for a perfect gate operation. The parameter variances are

$$\Delta\Psi = \Psi - \pi = \Phi_{\uparrow\uparrow} + \Phi_{\downarrow\downarrow} - (\Phi_{\uparrow\downarrow} + \Phi_{\downarrow\uparrow}) - \pi \qquad (3.63)$$

$$\Delta\theta_j = \theta_j - \bar{\theta}_j = \Delta\theta_{\text{LS}} + \Delta\Theta$$

The mean single qubit phase $\bar{\theta}_j = \bar{\Theta}_j + \bar{\theta}_{\text{LS}} = \bar{\Theta}_j$, because $\bar{\theta}_{\text{LS}}$ has to be zero for minimal $\Delta\theta_{\text{LS}}^2 = 2|\vartheta^+|^2$. We can scale the Rabi frequency such that $\bar{\Psi} = \pi$, and therefore the remaining error of the two-qubit phase stems from the variance of the geometric phase due to the randomness of ϕ_0, i.e. $\Delta\Psi^2 \propto \Delta\Phi^2 = \text{Var}[\Phi] = |I^+ - I^{-*}|^2/2$. A summary of all relevant phases and their meanings can be found in Appendix C.4.

The error ϵ includes all conditions (i)-(iii) we have for a fast geometric phase gate robust to changes in ϕ_0. We can therefore find suitable gate sequences by minimizing ϵ. While ϵ is a good measure to efficiently characterise and optimise gate sequences, it is only a convex approximation to the full gate error. Full errors are calculated from the overlap of the fully numerically integrated state with the target state, including mixed error terms.

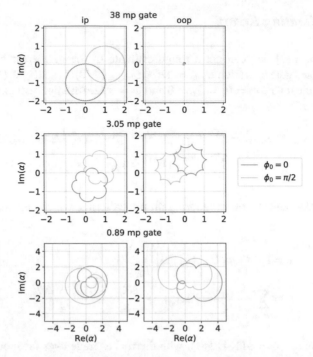

Fig. 3.6 Phase space trajectories in the rotating frame for example gate sequences in three different time regimes. For a conventional adiabatic gate (top) the beat note frequency is close to the ip mode and barely excites the oop mode. The phase-space trajectory is nearly circular and ϕ_0 affects the orientation of the circle, but not its shape or area. The centre plot shows a trajectory where the driving force couples comparably to both modes. Geometric phase is acquired from both modes and the shape of the trajectories changes slightly for different ϕ_0. Amplitude shaping is necessary to close the loops for both modes (as can be seen in the different sizes of the "petals") and to ensure the net geometric phase is constant for all ϕ_0. The bottom plot shows a trajectory below one ip motional period: the shape of the trajectory now depends strongly on ϕ_0. The plot shows a simulation within the Lamb-Dicke approximation; however, at this timescale out-of-Lamb-Dicke effects become important and mean that the loops no longer close for all ϕ_0, leading to significant gate errors, see Sect. 3.7.7. Figure published in [34]

3.6.3 Finding Appropriate Gate Sequences

Following [32], we select gate sequences by starting from a randomly seeded sequence, and then apply a Nelder-Mead optimiser that adjusts parameters to min-imise ϵ. A solution is chosen if the gate error $\epsilon < 10^{-4}$. These possible solutions were filtered further for minimum integrated and peak Rabi frequency (to minimise scattering errors). Amongst the remaining solutions sequences were chosen that did not have overly sharp features (i.e. many and/or large changes of pulse amplitude for short durations) to facilitate their implementation.

A numerical simulation of the gate dynamics in the rotating frame is shown in Fig. 3.6, where we characterise the gate duration in ip motional periods, with $1 \text{ mp} = \frac{1}{f_{z,\text{ip}}}$. We can distinguish four different regimes, see Fig. 3.7:

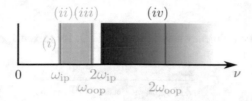

Fig. 3.7 Fast gate regimes: Depending on the Raman beatnote we can distinguish between (i) adiabatic gates, (ii) intermediate speed gates, (iii) efficient fast gates and (iv) highly non-adiabatic gates. Performing gates in the white areas is possible, but leads to larger errors than performing gates of similar speed at different detuning

(i) Adiabatic gates, where $|\nu - \omega_z| \ll \omega_z$. The rotating wave approximation is valid and gates can be performed on a single mode for arbitrary t_g or δ_g with rectangular pulses. To suppress off-resonant excitation the pulse edges are shaped with $t_{\text{rise}} \approx 1.5\,\mu s$.

(ii) Intermediate speed gates where $t_g \approx 4 - 10$ mp. There are good solutions for a single rectangular gate pulse with a specific detuning δ_g, that can for certain gate durations t_g close loops for both motional modes sufficiently well.

(iii) Efficient fast gates of $t_g \approx 2 - 5$ mp, where $\omega_{z,\text{ip}} < \nu < \omega_{z,\text{oop}}$, $|\nu - \omega_z| \sim \omega_z$. Only certain gate sequences with specifically timed pulses, shaped amplitudes and Raman beatnote ν yield low errors. Both motional modes are excited and contribute constructively to the geometric phase.

(iv) Strongly non-adiabatic fast gates with $\nu > \omega_{z,\text{oop}}$, $|\nu - \omega_z| \sim \omega_z$. Geometric phase is acquired from both motional modes, but with opposite sign, therefore cancelling partially. This means larger Rabi frequencies and larger excursions in motional phase space are necessary, requiring more laser power and making the gate more susceptible to various sources of error.

We investigate 2 different kinds of solutions: single rectangular pulses for intermediate speed gates in regime (ii), and 7-segment symmetric sequences for fast gates in regimes (iii) and (iv). For the 7-segment sequences each segment of duration t_i has a constant amplitude a_i and the parameters of segments 1&7, 2&6 and 3&5 are identical, see Fig. 7.2, i.e. the sequence is symmetric in time. A 'binary' variation of the 7-segment sequence was also tested, with $a_1 = a_3 = a_5 = a_7 = 1$ and $a_2 = a_4 = a_6 = 0$. For gates in regime (iii) there are also sequences with only 5 segments that have sufficiently low errors. Sequences for all these kinds of solutions were experimentally implemented and characterised. We have also numerically tested sequences with a larger number of segments or different amplitude shaping, e.g. a smooth Gaussian amplitude, but these showed no advantages in fidelity. Therefore only the experimentally easier 'step-shaped' sequences were implemented.

Gate Efficiency

We can define a gate efficiency similar to Eq. (3.53) for mixed species gates with

$$\tilde{\zeta} = \frac{\sum_{m=\{ip,oop\}} \Phi_{odd,m} - \Phi_{even,m}}{\sum_{m=\{ip,oop\}} \Phi_{odd,m} + \Phi_{even,m}} \tag{3.64}$$

Both ζ and $\tilde{\zeta}$ correlate to the magnitude of the Rabi frequency integrated over the gate duration. However ζ for mixed species scales with the asymmetry of the Rabi frequencies on $|\uparrow\rangle$ and $|\downarrow\rangle$. In contrast, $\tilde{\zeta}$ for fast gates[9] depends on the gate detuning that influences how asymmetrically the ip and oop mode are excited and whether the geometric phases of the two modes add or subtract. For gates in regime (i) the efficiency $\tilde{\zeta} \approx 1$. For gates in regime (ii) and (iii) we have $\tilde{\zeta} > 1$, and for most gates in regime (iv) $\tilde{\zeta} < 1$. Gates with higher efficiency have a lower Raman scattering error, as the power is used more effectively.

3.7   Two-Qubit Gate Errors

The sources of error in the σ_z geometric phase gate were studied in detail in [5]. We revisit here the dominant sources of error and consider how they need to be extended for mixed-species and fast gates. We also look at additional error sources that only become relevant for fast gates. For modelling the errors of adiabatic gates we use the QuTiP python package [35] and numerically integrate the Lindblad master equation

$$\dot{\rho}(t) = -\frac{i}{\hbar}[H(t), \rho(t)] + \sum_n \frac{1}{2} \left[2L_n \rho(t) L_n^\dagger - \rho(t) L_n^\dagger L_n - L_n^\dagger L_n \rho(t) \right] \tag{3.65}$$

with $H(t)$ the gate Hamiltonian 3.48 and Lindblad collapse operators L_n. For imprecisely set gate parameters we can find analytic solutions modelling the gate errors. For fast gates error terms are added to the integrator used for finding gate sequences.

We define the error $\epsilon = 1 - \mathscr{F}$ using the square fidelity [4] which prevails in experimental work. For a pure target state

$$\mathscr{F} = \langle \Phi | \rho | \Phi \rangle = \mathrm{tr}(\sigma \rho) \tag{3.66}$$

Here $|\Phi\rangle = \frac{1}{\sqrt{2}}(|\uparrow\uparrow\rangle - i|\downarrow\downarrow\rangle)$ and $\sigma = |\Phi\rangle\langle\Phi|$. This is the square of the QuTiP standard implemented fidelity $\mathscr{F}_{QuTiP} = \sqrt{\langle \Phi | \rho | \Phi \rangle}$.

---

[9]Note that we cannot use this gate efficiency for the error calculations we will derive in Sects. 3.7.2 and 3.7.3. The simulations in those sections assume the rotating wave approximation and excitation of a single mode only. The parameter sensitivity is therefore less complex than for fast gates.

For mixed species gates the main difference compared with same species gates is the reduction of gate efficiency, which increases sensitivity to motional errors and imperfectly closed loops.

3.7.1 Photon Scattering

One of the largest, and the only fundamentally limited error source in two-qubit gates, is photon scattering, see Sect. 3.2.4. We model the errors caused by photon scattering with their corresponding Lindblad operators.

Raman Scattering

We model Raman scattering with the 4 Lindblad operators $L_n = \sqrt{\Gamma_{\text{Raman},\uparrow/\downarrow}}\sigma_{\pm,1/2}$ [5, 18], where $\sigma_{+,1} = \sigma_+ \otimes I_2 \otimes I_{n_{\text{HO}}}$, $\sigma_{-,2} = I_2 \otimes \sigma_- \otimes I_{n_{\text{HO}}}$. Here n_{HO} is the modelled dimension of the motional Hilbert space, with $n_{\text{HO}} = 15$ for all simulations in this chapter. In the absence of a gate Hamiltonian these Lindblad operators lead to a population decay $p_\downarrow(t) = \frac{1}{\Gamma_{\text{Raman},\uparrow}+\Gamma_{\text{Raman},\downarrow}}\left(\Gamma_{\text{Raman},\uparrow} + \Gamma_{\text{Raman},\downarrow}e^{-(\Gamma_{\text{Raman},\uparrow}+\Gamma_{\text{Raman},\downarrow})t}\right)$ for a single ion. For a σ_z two qubit geometric phase gate the error amounts to

$$\epsilon_{\text{Raman}} = \frac{3}{4}(\Gamma_{\text{Raman},\uparrow} + \Gamma_{\text{Raman},\downarrow})t_{\text{gt}} \tag{3.67}$$

where the total gate time $t_{\text{gt}} = Kt_g$, and K is the number of loops. In this calculation we have assumed that all Raman scattering events finish in the opposite qubit state. However $\Gamma_{\text{Raman},\uparrow/\downarrow}$ also includes scattering into other states in the ground state manifold. This will lead to slight changes in the gate error.

For scattering into the D states the electron leaves the qubit manifold entirely and the error on the gate is [5]

$$\epsilon_D = 2B_{r,D}\Gamma_{tot}t_{\text{gt}} \tag{3.68}$$

For mixed species and most fast gates the gate efficiency $\zeta < 1$, i.e. more integrated Rabi frequency is required to obtain a certain geometric phase compared to a standard gate with $\zeta = 1$. The Raman scattering rate is proportional to g^2. Therefore the scattering error for less efficient gates is larger than for a standard gate of equal length t_{gt}.

Rayleigh Dephasing

Errors due to Rayleigh dephasing can be modelled with $L_n = \sqrt{\Gamma_{el}/4}\,\sigma_{z,1/2}$ [5, 18]. Applying L_n inside a Ramsey interferometer on a single ion leads to decay of the Ramsey contrast of $e^{-\Gamma t/2}$. The integrated error for a two-qubit gate is

$$\epsilon_{\text{Rayleigh}} = \frac{\Gamma_{el}t_{\text{gt}}}{2} \tag{3.69}$$

Again the error due to Rayleigh dephasing increases for less efficient gates due to an increase of Γ_{el}.

3.7.2   Motional Errors

Although the σ_z geometric phase gate is—within the Lamb-Dicke approximation—insensitive to the ions' temperature, heating processes during the gate do cause errors. Heating can be caused by fluctuations of the trapping voltages, however the observed heating is much larger than can be explained by Johnson thermal noise [36]. The origin of the remaining 'anomalous' heating is still not understood to date, however it can be reduced by cooling the trap to cryogenic temperatures [37], increasing the ion-trap electrode distance [38], or in in-situ cleaning of the trap electrodes [39, 40]. Theories for possible explanations include fluctuating patch potentials on the trap electrodes [4]. While the underlying cause of heating is still wrapped in mystery, its phenomenological effect is easily measurable in the linear heating rate $\dot{\bar{n}}$. Heating can be modelled by coupling to a thermal amplitude reservoir [41]. Another motional effect is dephasing, which can be modelled by coupling to a thermal phase reservoir [41]. Amplitude heating always gives rise to motional dephasing, but is not necessarily the dominant source.

Motional Amplitude Heating

Coupling to a thermal amplitude reservoir can be modelled with the Lindblad operators $L_n = \sqrt{\gamma}a$, $\sqrt{\gamma}a^\dagger$ [5, 41] where the heating rate $\gamma = \dot{\bar{n}}$ causes a linear increase of a single ion's temperature with gradient $\dot{\bar{n}}$. This leads to a gate error

$$\epsilon_h = \frac{\gamma t_{gt}}{2K\zeta} \tag{3.70}$$

The error is larger for less efficient gates, because they entail a larger excursion in phase space and displacement of all spin states. This means there is larger spin-motion entanglement and therefore a change in the motional state will cause a larger error. For the same reason larger K reduces the gate error.

Motional Dephasing

Similar to photon scattering, heating processes that do not change the ion's state cause errors by decoherence. This can be modelled by coupling to a phase reservoir with Lindblad operator $L = \left(\sqrt{2/\tau}\right) a^\dagger a$ [5, 41]. Here τ is the motional coherence time between the two states $|n\rangle = 0$ and $|n\rangle = 1$. For a motional superposition state $|0\rangle + |1\rangle$ this leads to a decay of Ramsey contrast with $e^{-t/\tau}$. The motional coherence time can be measured experimentally, see Sect. 5.6.2. The integrated gate error due to motional decoherence is

$$\epsilon_{md} = \frac{\alpha_K t_{gt}}{\tau \zeta^z} \tag{3.71}$$

While for $\tau \to \infty$ the exponent $z \to 1$, smaller motional coherence times reduce the influence of the gate efficiency and lead to smaller z. The linear factor α_K decreases with the number of loops K and also changes slightly with the motional dephasing time τ. For $\tau \to \infty$ we obtain $\alpha_2 = 0.297$, $\alpha_1 = 0.686$.

3.7.3 Gate Parameter Imprecision

There are three parameters that have to be set interdependently for a two-qubit gate operation: the gate detuning δ_g, the gate duration t_g and the Rabi frequency Ω_R. A mis-set in δ_g or t_g will cause imperfect closure of the loops in phase-space, leading to residual spin-motion entanglement, and a mis-set δ_g will additionally cause a small error of the geometric phase. An incorrect Rabi frequency only changes the size of the loop and therefore only causes an error in the acquired geometric phase. For all three parameters the gate error scales quadratically with the relative parameter-offset in the limit of small parameter mis-sets.

3.7.3.1 Gate Detuning

We model incorrectly set gate detuning by numerically integrating the gate Hamiltonian for different offsets of δ_g. For small inaccuracies $\Delta \delta_g$ of the detuning the gate error scales as

$$\epsilon_{\delta_g} = \left(\frac{\Delta \delta_g}{\delta_g} \right)^2 \frac{\alpha}{\zeta^z} \tag{3.72}$$

For a two-loop gate we obtain in the limit of small errors $\alpha = \frac{5}{4}\pi^2 + 2\pi^2 \bar{n}$. For $|\zeta - 1| \ll 1$ we find $z = 0.81$, however for less efficient gates the description with $1/\zeta^z$-scaling starts to fail. For $0.5 < \zeta < 1$ the error can be described well with $z \approx 0.84$. The error increases for less efficient gates because the remaining spin-motion entanglement for a not fully-closed loop increases due to the larger overall-size of the loop in phase-space.

3.7.3.2 Gate Length

The geometric phase is in first order independent of mis-sets in t_g, leading to a slightly smaller relative gate error of

$$\epsilon_{t_g} = \left(\frac{\Delta t_{\text{gt}}}{t_{\text{gt}}}\right)^2 \frac{\alpha}{\zeta} \tag{3.73}$$

with $\alpha = \pi^2 + 2\pi^2 \bar{n}$ for a two-loop gate in the limit of small errors.

3.7.3.3 Rabi Frequency

Incorrectly set Rabi frequency does not affect loop closure and the error is therefore independent of the gate efficiency, the ions' temperature and the number of loops. The gate error scales like

$$\epsilon_{\Omega_R} = \left(\frac{\Delta \Omega_R}{\Omega_R}\right)^2 \alpha \tag{3.74}$$

where $\alpha = \frac{\pi^2}{4}$ in the limit of small errors.

3.7.3.4 Fast Gates

For fast gates the gate error scales quadratically for small errors in the pulse durations t_i, pulse amplitudes a_i, as well as the gate detuning δ_g, and the single qubit phase ϕ_{sq}. The coefficients for each parameter depend on the individual gate sequence. Gate errors were first simulated with a Monte-Carlo simulation, randomly adding normally distributed errors to all parameters. The error was then calculated by numerically integrating a pulse sequence with these parameters and calculating the overlap with the target state. For comparison a large set of errors for the t_i and a_i parameters was measured from real sequences. The gate error was then calculated from the measured noisy parameters and agreed well with the Monte Carlo simulations. Example simulations of linear 2D-parameter scans are shown in Fig. 3.8.

3.7.4 Off-Resonant Excitation

Far-detuned coupling to other features in the frequency spectrum leads to additional undesired dynamics and therefore reduces the gate fidelity. For a gate on the mode $m = \{\text{ip, oop}\}$ other features are, see Fig. 3.9, (i) the carrier at $\nu = 0$, (ii) the other axial mode at $\nu = \omega_{z,m'}$, (iii) the opposite sideband at $\nu = -\omega_{z,m}$ and (iv) higher order sidebands at $\nu = \pm n\omega_{z,m}$.

For the σ_z phase gate coupling to the carrier does not drive spin flips, because there is no qubit frequency splitting between the two gate lasers. Instead coupling to the carrier is in the form of a light shift that gives rise to a single qubit phase, see Sect. 3.6.1.1. The error due to this phase is calculated for fast gates in Sect. 3.6.2. It

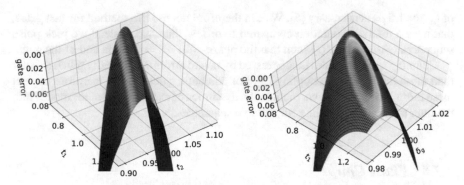

Fig. 3.8 Fast gate error simulations: Left: The sensitivity of the gate error to the pulse duration varies for different segments. There is typically a fairly long ridge, where pulse lengths can be traded off each other without a significant loss in fidelity. The scanned parameters are plotted in relative units, i.e. $t_1 = \Delta t_1 / t_{1_0}$, where t_{1_0} is the ideal pulse duration predicted by simulations. Right: The gate detuning is the most sensitive parameter in relative units, and can be typically barely compensated for by adjusting other parameters

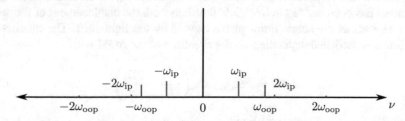

Fig. 3.9 Mode spectrum: The first sideband at $\pm \omega_m$ is suppressed by η_m relative to the carrier, the second sideband at $\pm 2\omega_m$ is suppressed by η_m^2. The relative frequencies of the two axial modes are given by $\omega_{\text{oop}} = \sqrt{3}\omega_{\text{ip}}$

is the same for adiabatic gates [5], and for a rectangular pulse of constant amplitude we obtain

$$\epsilon_{\text{LS}} = \frac{\Delta\theta_1^2 + \Delta\theta_2^2}{4} = |\vartheta^+|^2$$

$$= \left(\frac{\Omega_\uparrow - \Omega_\downarrow}{\nu}\right)^2 \sin^2\left(\frac{\nu t_g}{2}\right) \tag{3.75}$$

For adiabatic gates ($|\nu - \omega_m| \ll \omega_m$) this would give a maximum error[10] of $\epsilon_{\text{LS}} \sim 10^{-2}$ and for fast gates the error would be overwhelming. We can suppress this error by smoothing the pulse edges, for example with a $\sin^2(t)$-window. This removes higher frequency components and therefore greatly reduces off-resonant coupling. To sufficiently suppress errors to a negligible level ($\epsilon_{\text{LS}} \sim 10^{-5}$) a shaping duration

---

[10]Assuming $\nu \approx 2\pi \cdot 2\,\text{MHz}$ and $\Delta\Omega \approx 2\pi \cdot 200\,\text{kHz}$ for a typical adiabatic gate operation.

of $t_{\text{rise}} \approx 1.5\,\mu$s is necessary [5]. We can therefore not use this method for fast gates, that have a total gate duration compared to or less than t_{rise}. Instead we pick pulse sequences with t_g and $\Omega(t)$ such that the phase-shift $\vartheta^+ = 0$ at the end of the gate.

Coupling to other modes is suppressed by η, and errors are typically much smaller than those due to carrier coupling [5]. For adiabatic gates we can therefore neglect these terms. For fast gates we include the dynamics of excitation of other modes into the integrator when searching for suitable sequences.

3.7.5 Phase Chirps

In contrast to slow (adiabatic) gates, fast gates are much more sensitive to phase fluctuations within the laser pulses due to the short length of the pulses. The on- and off-switching of the laser pulses is performed with acousto-optic modulators (AOMs), that can cause phase chirps during the switching [42]. We model the phase chirp with a time-dependent quadrature field amplitude $\Omega_q(t)$. Its effect on the gate dynamics has been derived in Sect. 3.6. It affects both the displacement of the spin-states as well as the single qubit phase caused by the light-shift. The quadrature amplitude is modelled depending on the rise-time of the AOM with

$$\Omega_q(t) = \beta \Omega_{i,\max} \sin\left(\frac{1}{\Omega_{i,\max}} \frac{d\Omega_i}{dt}\right) \tag{3.76}$$

where β is a scaling factor inferred from experimental measurements of the AOM's phase chirp. The gate error is then determined by integrating the gate dynamics for a given β. For small errors the gate error scales with $\epsilon_{\text{pc}} = \alpha\beta^2$, where the proportionality factor α increases for shorter gate sequences. For example $\alpha = 1.1$ for a 0.89 mp (motional periods of the ip axial mode) gate sequence and $\alpha = 0.05$ for a 3.05 mp gate sequence.

3.7.6 Radial Mode Coupling

For fast gates the laser beatnote can be close to resonance with radial mode frequencies. If the differential Δk vector of the Raman lasers is not perfectly parallel to the trap axis this can cause excitation of the radial mode motion. However gate sequences are not designed to close radial mode loops, and the remaining spin-motion entanglement will cause errors. Because the radial modes are only cooled with Doppler/ dark resonance cooling they are at a higher temperature than the axial modes, further increasing gate errors. Because of the orthogonality of the modes we can model errors due to radial mode coupling by calculating the gate fidelity of a sequence applied to the four radial modes only, with the corresponding frequencies, mode temperatures and Lamb-Dicke factors. The Lamb-Dicke parameter depends on the angle between the trap axis and Δk. Errors due to radial mode coupling strongly depend on the Raman beatnote ν. Results of the simulation are shown in Sect. 7.3.3.

3.7.7 Out-of-Lamb-Dicke Effects

While the Lamb-Dicke approximation is fulfilled fairly well for adiabatic gates, fast gates require larger displacements in phase space due to their lower gate efficiency, and thus involve considerable deviations from the Lamb-Dicke regime. The force exerted by the light field is no longer position independent, and therefore not only leads to a displacement in phase space, but also to squeezing terms. We further have to take into account that the Rabi frequency decreases with larger motional mode occupation n and therefore for a thermal state is inhomogeneous. The relative fluctuations of the absolute Rabi frequency due to a thermally occupied spectator mode p can be expressed as [4]

$$\left(\frac{\Delta\Omega_R}{\bar{\Omega}_R}\right)^2 \approx \eta_p^4 \bar{n}_p (\bar{n}_p + 1) \tag{3.77}$$

leading to a gate error $\epsilon_{\Omega_R} \approx \frac{\pi^2}{4} \eta_p^4 \bar{n}_p (\bar{n}_p + 1)$.

The out-of-Lamb-Dicke (ooLD) effects caused by thermal occupation and displacement of the gate mode are harder to model analytically. Instead, we perform a numerical simulation of the full gate Hamiltonian without Lamb-Dicke approximation. The Hamiltonian is integrated using the split-operator method, and the gate error is averaged over different initial phases of the Raman beams ϕ_0. Outside of the Lamb-Dicke regime the gate sequences are no longer insensitive to ϕ_0. This is because for different initial phases large displacements in phase space—and therefore stronger ooLD modulation—occur at different times in the gate sequence. This leads to substantial errors. Squeezing terms lead to additional errors.

Due to its slow performance the ooLD integrator was only used to re-optimise sequences formerly found within the Lamb-Dicke approximation. Solutions of the simulations are shown in Sect. 7.3.4. For gate sequences below $\approx$2.5 mp ooLD errors are the dominant source of error.

References

1. Kotler S, Akerman N, Navon N, Glickman Y, Ozeri R (2014) Measurement of the magnetic interaction between two bound electrons of two separate ions. Nature 510:376–380. ISSN: 1476-4687
2. Dehmelt HG (1968) In: Advances in atomic and molecular physics, pp 53–72
3. Paul W, Steinwedel H (1953) Ein neues Massenspektrometer ohne Magnetfeld. Zeitschrift für Naturforschung A 8:448–450

4. Wineland DJ et al (1998) Experimental issues in coherent quantum-state manipulation of trapped atomic ions. J Res Nat Inst Stand Technol 103:259–328
5. Ballance CJ (2014) High-fidelity quantum logic in Ca+. PhD thesis, University of Oxford
6. Leibfried D, Blatt R, Monroe C, Wineland D (2003) Quantum dynamics of single trapped ions. Rev Mod Phys 75:281–324. ISSN: 00346861
7. Berkeland DJ, Miller JD, Bergquist JC, Itano WM, Wineland DJ (1998) Minimization of ion micromotion in a Paul trap. J Appl Phys 83:5025–5033. ISSN: 00218979
8. Wineland DJ et al (1998) Experimental primer on the trapped ion quantum computer. Fortschritte der Physik 46:363–390. ISSN: 0015-8208
9. Wübbena JB, Amairi S, Mandel O, Schmidt PO (2012) Sympathetic cooling of mixed-species two-ion crystals for precision spectroscopy. Phys Rev A 85:043412
10. Haroche S, Raimond J-M (2006) Exploring the quantum. Oxford University Press, London, UK
11. Foot CJ (2005) Atomic physics. Oxford University Press, London, UK
12. Scully MO, Zubairy MS (2002) Quantum optics. Cambridge University Press, Cambridge, UK
13. Loudon R (1983) The quantum theory of light. Oxford University Press, London, UK
14. Wineland DJ et al (2003) Quantum information processing with trapped ions. Philos Trans R Soc London A 361:1349–1361. ISSN: 1364-503X
15. Steck DA (2007) Quantum and atom optics. Tech Rep, University of Oregon
16. James DFV, Jerke J (2007) Effective Hamiltonian theory and its applications in quantum information. Can J Phys 85:625–632
17. Ozeri R et al (2005) Hyperfine coherence in the presence of spontaneous photon scattering. Phys Rev Lett 95:030403. ISSN: 0031-9007
18. Uys H et al (2010) Decoherence due to elastic Rayleigh scattering. Phys Rev Lett 105:200401. ISSN: 00319007
19. Ozeri R et al (2007) Errors in trapped-ion quantum gates due to spontaneous photon scattering. Phys Rev A 75:042329. ISSN: 1050-2947
20. Langer CE (2006) High fidelity quantum information processing with trapped ions. PhD thesis, University of Colorado
21. Cirac JI, Zoller P (1995) Quantum computations with cold trapped ions. Phys Rev Lett 74:4091–4094. ISSN: 00319007
22. Sørensen A, Mølmer K (1999) Quantum computation with ions in thermal motion. Phys Rev Lett 82:1971–1974. ISSN: 0031-9007
23. Leibfried D et al (2003) Experimental demonstration of a robust, high-fidelity geometric two ion-qubit phase gate. Nature 422:412–415. ISSN: 0028-0836
24. Bermudez A, Schmidt PO, Plenio MB, Retzker A (2012) Robust trappedion quantum logic gates by continuous dynamical decoupling. Phys Rev A 85:040302. ISSN: 10502947
25. Lee PJ et al (2005) Phase control of trapped ion quantum gates. J Opt B: Quantum Semi-classl Opt 7:371–383. ISSN: 1464-4266
26. Carruthers P, Nieto MM (1965) Coherent states and the forced quantum oscillator. Am J Phys 33:537–544
27. Ozeri R (2011) The trapped-ion qubit tool box. Contemp Phys 52:531–550. ISSN: 0010-7514
28. Home JP (2013) Quantum science and metrology with mixed-species ion chains. Adv At Mol Opt Phys 62:231–277. ISSN: 1049250X
29. García-Ripoll JJ, Zoller P, Cirac JI (2003) Speed optimized two-qubit gates with laser coherent control techniques for ion trap quantum computing. Phys Rev Lett 91:157901. ISSN: 0031-9007
30. Duan L-M (2004) Scaling ion trap quantum computation through fast quantum gates. Phys Rev Lett 93:100502
31. Bentley CDB, Carvalho ARR, Kielpinski D, Hope J (2013) Fast gates for ion traps by splitting laser pulses. New J Phys 15. ISSN: 13672630
32. Steane AM, Imreh G, Home JP, Leibfried D (2014) Pulsed force sequences for fast phase-insensitive quantum gates in trapped ions. New J Phys 16. ISSN: 13672630

33. Palmero M, Martinez-Garaot S, Leibfried D, Wineland DJ, Muga JG (2017) Fast phase gates with trapped ions. Phys Rev A 95:022328
34. Schäfer VM et al (2017) Fast quantum logic gates with trapped-ion qubits. Nature 555:75–78. ISSN: 0028-0836 (2017)
35. Johansson JR, Nation PD, Nori F (2013) QuTiP 2: a Python framework for the dynamics of open quantum systems. Comp Phys Comm 184
36. Turchette QA et al (2000) Heating of trapped ions from the quantum ground state. Phys Rev A 61:063418. ISSN: 1050-2947
37. Deslauriers L et al (2006) Scaling and suppression of anomalous heating in ion traps. Phys Rev Lett 97:103007. ISSN: 00319007
38. Sedlacek JA et al (2018) Distance scaling of electric-field noise in a surface electrode ion trap. Phys Rev A 97:020302. ISSN: 24699934
39. Allcock DTC et al (2011) Reduction of heating rate in a microfabricated ion trap by pulsed-laser cleaning. New J Phys 13. ISSN: 13672630
40. McConnell R, Bruzewicz C, Chiaverini J, Sage J (2015) Reduction of trapped ion anomalous heating by in situ surface plasma cleaning. Phys Rev A 92:020302. ISSN: 10941622
41. Turchette QA et al (2000) Decoherence and decay of motional quantum states of a trapped atom coupled to engineered reservoirs. Phys Rev A 62:053807. ISSN: 10502947
42. Degenhardt C et al (2005) Influence of chirped excitation pulses in an optical clock with ultracold calcium atoms. IEEE Trans Instrum Meas 54:771–775. ISSN: 00189456

Chapter 4
Apparatus

An ion trap experiment consists of the trap and vacuum system, as well as systems to control the ions and the environment. The ion trap is housed in a vacuum chamber, surrounded by coils to produce and orient the desired magnetic field. For controlling the ions we use multiple lasers as well as microwaves and RF. Information is obtained from the ions by measuring their fluorescence, detected by either an EMCCD camera, or an APD detector. An overview of the entire experimental apparatus is shown in Fig. 4.1.

4.1 The Ion Trap

The ion trap is a linear Paul blade trap, designed by Sarah Woodrow and built by Keshav Thirumalai. A sketch of the design is shown in Fig. 4.2 and details about the design process can be found in [1]. The trap was designed to have a moderately large micromotion gradient along the trap axis $\hat{z}$ to be able to address ions using the micromotion sideband. For this purpose the ion-endcap distance is relatively short with $d = 1.15$ mm. The ion-blade distance is $\rho = 0.5$ mm. Simulated geometry factors are (for asymmetric driving, i.e. with two RF blades grounded) $\alpha = 0.207$, $\frac{\kappa_x}{2} = 0.486$, $\frac{\kappa_y}{2} = 0.487$ and $\kappa_z = 0.104 = \frac{\alpha}{2}$ [1].

The ion trap is inside a vacuum system built by Keshav Thirumalai and depicted in Fig. 4.3. The system is operated at room temperature and the operating pressure is off-scale low on the ion gauge ($<1 \times 10^{-11}$ Torr). There are two ovens in the system—one containing strontium for loading ^{88}Sr$^+$, and one containing enriched[1] calcium for loading ^{43}Ca$^+$ and ^{40}Ca$^+$. The ovens are operated by applying an $I \approx 4$ A current, which causes some of the contents to evaporate. The oven currents are adjusted for convenient loading rate.

---

[1] 12% ^{43}Ca$^+$, 88% ^{40}Ca$^+$.

© Springer Nature Switzerland AG 2020
V. M. Schäfer, *Fast Gates and Mixed-Species Entanglement with Trapped Ions*,
Springer Theses, https://doi.org/10.1007/978-3-030-40285-3_4

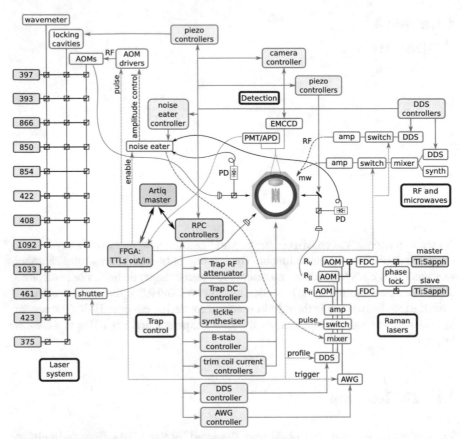

Fig. 4.1 Apparatus: All experimental control is managed by the Artiq master. Software RPC controllers (light green boxes) run different bits of physical hardware (grey-lined white boxes) and are themselves controlled by the master. The Artiq FPGA board creates real-time pulse sequences and sends out TTL pulses to trigger the various devices. Solid green lines represent optical fibres, and dashed green lines represent SMA cables. Abbreviations used are acousto optic modulator (AOM), radio frequency radiation (RF), direct current (DC), electron multiplying charge coupled device (EMCCD), photon multiplier tube (PMT), avalanche photodiode (APD), photodiode (PD), direct digital synthesis (DDS), synthesiser (synth), microwaves (mw), frequency doubling cavity (FDC), amplifier (amp), B-field stabilisation (B-stab), field programmable gate array (FPGA), transisitor-transistor logic (TTL), remote procedure call (RPC) and arbitrary waveform generator (AWG)

4.1.1 Trap Voltages

The voltages on the trap electrodes can be adjusted for different trap depths and frequencies. We use a weaker trap (WT) for loading, a tight trap (TT) for all experiments and an intermediate trap (IT) when switching between the two. The different electrodes are the endcaps (far and near, FEC and NEC), compensation electrodes

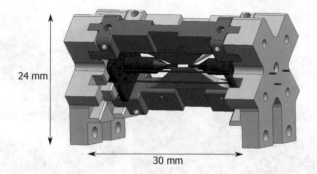

Fig. 4.2 The ion trap: Radial confinement is provided by the trap RF, which is applied to one pair of the blades (blue), while the other pair is kept at ground level (yellow). The second grounded blade is not shown for better visibility of the trap centre. DC voltage applied to the endcaps (red) produces a harmonic potential in the axial direction. Compensation electrodes (purple) are used to correct for machining imperfections and place the ion at the center of the RF trap potential. The trap electrodes are held in place by insulating Macor ceramic (grey). Figure from [1]

(bottom, far and near, BC, FC and NC) and RF blades, where the RF voltage is tuned with a variable attenuator. The applied voltages are listed in Table 4.1.

Trap RF

The trap RF is produced by a synthesiser[2] and the signal is amplified[3] and fed to a helical resonator, which transforms the large current signal into a high voltage signal. The helical resonator is of similar design to [3], with one end of the helix connected to the two driven RF blades via a feed-through and the other end connected to the grounded blades via the vacuum chamber. A small coil at the grounded end is used to couple the RF into the resonator. The loaded quality factor of the resonator is $Q \approx 300$ and its resonance frequency is $f_{RF} = 28.0133$ MHz at axial frequency $f_z = 1.860$ MHz, leading to radial frequencies $f_{rad,lower} = 4.077$ MHz and $f_{rad,upper} = 4.341$ MHz. A digital step attenuator[4] in the RF supply chain is used to adjust the RF amplitude to vary between different trap depths.

Trap DC

The trap DC supply is a custom design by Chris Ballance and described in [3]. The output voltage is -240 V $< U_{DC} < +240$ V with a spectral noise density $\ll$ $1 \, \mu V / \sqrt{Hz}$ around $f = 2$ MHz. A filter board directly before the feed-through to the trap filters out any noise picked up in the cables. An additional connection bypassing the filter board allows application of an AC signal 'tickle' onto one of the endcaps

---

[2]Hewlett Packard, 8656B, 6.1–990 MHz.

[3]Frankonia, FLL-25, 100 kHz–250 MHz, +46 dB, up to 25 W output power.

[4]Mini circuits, ZX76-31R5-PP+, DC-2400 MHz, 31.5 dB maximum attenuation with 0.5 dB steps, 6 bit interface.

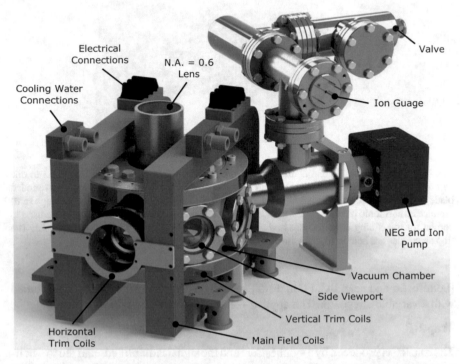

Fig. 4.3 Vacuum system: The ion trap is housed in an octagonal vacuum chamber with 6 fused silica viewports on the sides and one recessed viewport on top. High-field magnetic coils designed to produce fields up to 288 G give access to multiple clock transitions in ^{43}Ca$^+$. Two pairs of horizontal and vertical trim coils can be used to superimpose the magnetic field axis with the 397σ laser beam, for optimal polarisation. Imaging of the ion is performed from top, with a high NA = 0.6 lens. Electrical feed-throughs are on the bottom plate. A combined non-evaporable getter (NEG) and ion pump are used to maintain the vacuum. Figure from [2]

Table 4.1 Trap voltages applied for different trap depths. The voltages of the compensation electrodes had to be adjusted regularly due to drifts. The voltages given are the ones used during the fast gates measurements

	RF (dB)	U_{NEC} (V)	U_{FEC} (V)	U_{NC} (V)	U_{FC} (V)	U_{BC} (V)
TT	−3.5	191.14	193.27	−50	50	−60
IT	−3.5	19.9	20.0	−50	50	−60
WT	−14	5.74	6.01	−2	2	−150

or one of the compensation electrodes. The AC signal can be used to excite the ion's motion if it is resonant with a trap frequency. This modifies fluorescence and can therefore be used to determine the trap frequencies.

4.1.2 Imaging System

The imaging system was designed and built by Chris Ballance. The ions are imaged with a high-NA lens[5] with NA $= 0.6$ towards the atom and NA $= 0.09$ on the imaging side, leading to a magnification of $\times 6.67$. The beam is collimated using a $f = 20$ mm aspheric lens,[6] and then either focussed onto an EMCCD camera[7] with a $f = 150$ mm lens, or onto an APD[8] with a $f = 11$ mm lens. The high-NA lens is suspended from and aligned with a 3D translation stage[9] in combination with two rotation stages.[10] The total magnification onto the camera is $\times 50.3$ and thus two ions spaced apart by $d = 3.5 \, \mu$m are separated by 11 pixels. To reduce scatter, a rectangular aperture that is adjustable in both dimensions is placed in the focal plane of the high-NA lens. An additional iris between the high-NA lens and its focus can be used to reduce the effective NA and ease alignment. A mirror on a motorised[11] mount is used to toggle between the camera and the APD.

4.2 Lasers and Beam Geometry

All calcium lasers, together with their locking cavities and AOM networks are on a separate optics table, the laser table. The beams are then coupled into optical fibres and delivered to the trap table. On the trap table there is only a short beam path for polarisation and intensity stabilisation before the beams are focussed onto the ions. This separation improves beam pointing stability and the beam profile at the trap. The strontium lasers and AOMs are on breadboards mounted in a rack [2, 4], with only the optical locking cavities still mounted on an optical table. The strontium laser setup was built by Keshav Thirumalai.

4.2.1 Laser Table

All lasers apart from the Raman lasers are external cavity diode lasers[12] with Faraday isolators integrated into the laser head. For cooling and SPAM we have 5 lasers for calcium (397, 393, 866, 850 and 854 nm) and 4 lasers for strontium (422, 408,

---

[5]Photon gear inc., Atom Imager Relay, 422 nm, PGI P/N 16580.

[6]Asphericon, AFL12-20-S-U.

[7]Andor Technologies, iXon^{EM} DU-897E, up to $\times 1000$ EM gain, image area 8.2 mm $\times$ 8.2 mm, pixel size 16 μm $\times$ 16 μm, 55% quantum efficiency at 400 nm.

[8]Micro Photon Devices PD-100-CTC.

[9]Newport, ULTRAlign M-562-XYZ.

[10]Newport, M-GON65-U, goniometric rotation stage.

[11]Actuonix, linear actuator.

[12]Toptica DLC DL pro, short term (5 μs) linewidth <100 kHz (red) to <2 MHz (blue).

Table 4.2 Laser detunings: The laser frequencies are the wavemeter readings when the respective laser is on resonance with the given ion at low field. The detunings are the typical laser operation frequencies for ^{43}Ca$^+$ (^{88}Sr$^+$) at high field (146 G) relative to the low field ^{40}Ca$^+$ (^{88}Sr$^+$) resonance. They consist of the Zeeman shift, the isotope and hyperfine shift (^{43}Ca$^+$ only) and any intentional detuning from resonance. The frequency at the ion is the sum of the laser frequency, the detuning and AOM shifts

Laser	Detuning	^{40}Ca$^+$ frequencies (THz)	Laser	Detuning	^{88}Sr$^+$ frequencies (THz)
397	-1.26 GHz	755.2223(2)	422	280 MHz	710.9626(2)
393	2.07 GHz	761.9046(2)	408	500 MHz	734.9899(2)
866	-3.49 GHz	346.0002(2)	1092	0 MHz	274.5890(2)
854	-4 GHz	350.8628(2)	1033	1 GHz	290.2106(2)
850	-3.12 GHz	352.6826(2)			
423	610 MHz	709.0781(2)	461	0 MHz	650.5039(2)

1092, 1033 nm). For species- and isotope-selective photo-ionisation we use a 423 nm laser for calcium and a 461 nm laser for strontium. The second step of photo-ionisation (PI) into the continuum is performed by a 375 nm laser[13] for both species. All lasers apart from the 375 are monitored on a wavemeter.[14] The laser light is guided to the wavemeter via multimode fibres, leading to drifts of $\approx \pm 150$ MHz on the wavelength reading. Typical wavemeter readings of all lasers are shown in Table 4.2. A fibre-beamsplitter before the wavemeter is used to guide light to diagnostic optical spectrum analyser.[15] The transmission signal of the spectrum analysers is monitored on a screen to detect multi-mode lasing. Apart from the PI lasers, the 1033 and the 854, all lasers are locked to low-drift reference cavities.[16] As the 408 laser is only used temporarily for readout until the 674 nm quadrupole laser is available, it is locked with a simple top-of-fringe lock. All remaining lasers are locked with Pound-Drever-Hall (PDH) locks,[17] where the sidebands are created with EOMs[18] for blue lasers and current modulation of the diode for red lasers. The 423 frequency is roughly stabilised with a wavemeter lock. The 397 and 866 lasers have a grating in the beginning of their beampath for filtering out amplified spontaneous emission (ASE). For the 397 laser sidebands at $f_{EOM} = 2.955$ GHz are added with an EOM[19]

---

[13]Toptica iBeam Smart.

[14]High Finesse, WS7 wavelength measuring system USB, 350–1120 nm, 150 MHz accuracy with multi-mode fibre input.

[15]Red lasers: Toptica, FPI-100-0980-1, 825–1200 nm, FSR = 1 GHz, , finesse $F \approx 500$ blue lasers: Thorlabs, SA200-3B, 350–535 nm, FSR = 1.5 GHz, finesse $F \approx 250$.

[16]Calcium: NPL Low Drift Etalon, 1.5 GHz FSR, specified drift <0.5 MHz/h strontium: Stable Laser systems, SLS-NPLcav-3, based on NPL cavity design, 1.5 GHz FSR, specified drift <0.5 MHz/h, Finesse = 60–180.

[17]Lock module: Toptica, PDD 110/F.

[18]QUBIG, EO-F80M3, 350–670 nm, MgO:LN.

[19]New Focus, 4435-02, 360–500 nm, 2.5–4.6 GHz, MgO:LiNbO$_3$.

to address population in both ground state manifolds of ^{43}Ca$^+$. The EOM frequency is optimised for the 146 G magnetic field and has to be adjusted for operation at different magnetic fields.

AOM networks

The laser beams are switched on and off[20] via AOMs in the beampath, with different types of AOMs used for blue lasers,[21] calcium red lasers[22] and strontium red lasers.[23] Only the beam diffracted by the AOM is coupled into the fibre leading to the trap.

The AOMs are also used for adding independent frequency offsets required for addressing different transitions. Beams are split up for the various AOMs using polarising beam splitters (PBSs) and the relative intensity is adjusted using $\lambda/2$ waveplates. The 397 laser is split into 3 beams. A 397σ and a 397π beam at equal frequency provide different polarisations and beam-directions for laser cooling and state preparation. A third 397_{load} beam is not diffracted by an AOM and only switched with a mechanical shutter. It is therefore detuned red by -400 MHz from the other two beams. It is applied onto the ion in a large spot size and mixed polarisation to aid cooling of hot ions during loading. The 866 laser is split into two beams 866 and 866_d differing by 270 MHz. The main 866 beam is used for high fluorescence Doppler cooling, while the 866_d beam is used only for dark resonance cooling. The 850 laser is split into 3 different beams addressing the different transitions $3D_{3/2}^{5,+5} \rightarrow 4P_{3/2}^{5,+5}$ (850π), $3D_{3/2}^{5,+4} \rightarrow 4P_{3/2}^{5,+5}$ ($850\sigma_1$) and $3D_{3/2}^{4,+4} \rightarrow 4P_{3/2}^{5,+5}$ ($850\sigma_2$) for recovering population from $3D_{3/2}$ during shelving. The 422 laser has beams of two different frequencies 422_u and 422_l to address population in both qubit levels and to allow frequency-selective state-preparation.

All AOMs are operated in double-pass geometry to improve extinction while the AOM is switched off. To improve extinction further, the calcium AOM drivers have a frequency offset added to their drive frequency whenever their amplitude is switched off. A summary of all different frequency offsets can be found in Table 4.3.

Beams of similar frequency and equal polarisation that enter the vacuum chamber from the same direction are superimposed on a beam-splitter before the fibre. A sequence of PBS-$\lambda/2$-PBS is used for optimal use of power of the respective lasers. The superimposed beams are (397σ, 393), (422_u, 422_l), (866, 866_d, 854), ($850\sigma_1$, $850\sigma_2$) and (1092, 1033).

---

[20] AOM driver: IntraAction DE-2001.5EM26A, also models with different frequency and output power.

[21] IntraAction Corp. ASM-2001.588.

[22] IntraAction ATM-851A2.

[23] Gooch & Housego, Fiber-Q, fibre coupled modulators (FCAOM).

Table 4.3 Beam intensities, polarisations $\hat{\epsilon}$ **and waist radii** r_0: Some beams have different power setpoints for fluorescence (F), dark resonance cooling (DRC), (low power) Doppler cooling (DC) and sideband cooling (SBC). The AOM frequency shifts are added after the wavemeter reading, and before the fibre leading to the trap table. Intensities are given in units of the saturation intensity I_0 for the relevant transition (defined in Eq. 4.1)

Laser	r_0 (μm)	Intensity	$\hat{\epsilon}$	AOM shift (MHz)
397π	45	$11I_0$ (F), $0.8I_0$ (DRC)	π	$+400$
397σ	43	$50\,I_0$ (F), $1.4\,I_0$ (DRC), $0.3\,I_0$ (SBC)	σ^+	$+400$
393	43	$0.017I_0$	σ^+	$+400$
866	112	$1300I_0$ (F), $60I_0$ (SBC)	$\sigma^\pm$	$+140$
866_{dark}	112	$250\,I_0$	$\sigma^\pm$	-130
854	112	$200\,I_0$	$\sigma^\pm$	$+170$
850π	112	$22\,I_0$	π	-208.5
$850\sigma_1$	114	$10\,I_0$	σ^+	-131.6
$850\sigma_2$	114	$10\,I_0$	σ^+	-361
422_u	65	$4\,I_0$ (F), $1I_0$ (DC)	$\sigma^\pm$	$+272$
422_l	65	$4\,I_0$ (F), $1I_0$ (DC)	$\sigma^\pm$	-272
408	65	$1.2\,I_0$	σ^+	$+400$
1092	112	$350\,I_0$	$\sigma^\pm$	$+300$
1033	112	$45\,I_0$	$\sigma^\pm$	$+300$

4.2.2 Trap Table

On the trap table a PBS is used to clean up the polarisation from shifts occurring in the fibre. The power of a pick-off of the beam is measured on a PD and the signal is used as input for a PID loop[24] to stabilise the intensity. The output signal is fed back to the AOM drivers to modulate the amplitude. The noise-eater has different power-setpoints that can be used to switch the power of a beam within a sequence.

Beams of orthogonal polarisation are then superimposed on PBSs, and beams of different frequencies are superimposed using dichroic filters/mirrors,[25] before they are focussed onto the ions. The polarisation of the 397σ, 393 and $850\sigma_{1,2}$ lasers is additionally improved with a Glan-laser polariser[26] and then transformed to circular polarisation using a $\lambda/4$ waveplate on a tip-tilt mount.

---

[24]Noise eater, Elephant electronics (i.e. Christopher Ballance and Thomas Harty), using MicroZed evaluation board.

[25]Thorlabs DMSP950 (950 nm shortpass), DMLP650 (650 nm longpass), Edmund Optics 86–323 (409 nm longpass).

[26]Thorlabs, GT10-A, Glan-Taylor Polariser, GT10-B for red lasers.

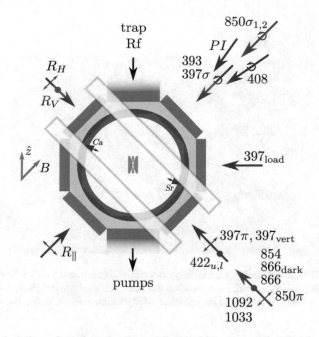

Fig. 4.4 Beam delivery into the trap: Beams displayed in columns or with different subscripts arrive in the same fibre. Other beams are superimposed with PBSs and dichroics. The magnetic field is aligned along the 397σ propagation axis using trim coils. The PI and 408 beams are at a slight angle because their polarisation purity is not as important. The 397π laser can be guided on an alternative path $10°$ out of plane with respect to the laser table, using a flip-mirror. This beam path is only used for vertical micromotion compensation of the trap. The calcium and strontium ovens are aligned such that the atomic beams are near-perpendicular to the PI beams, minimizing Doppler shifts

The alignment of the different beams into the trap is shown in Fig. 4.4. All beams are focussed with a $f = 400$ mm lens,[27] but have different spot-sizes due to different beam diameters caused by different fibre out-couplers. A summary of all different spot-sizes, as well as beam polarisations and power setpoints is shown in Table 4.3. The beam intensities are listed in units of I_0, where we use

$$I_0 = \frac{\hbar \omega_L^3}{6\pi c^2 \tau} \tag{4.1}$$

as definition for the saturation intensity [5]; note therefore that I_0 has a different value for each atomic transition.

---

[27]Thorlabs, AC254-400-A.

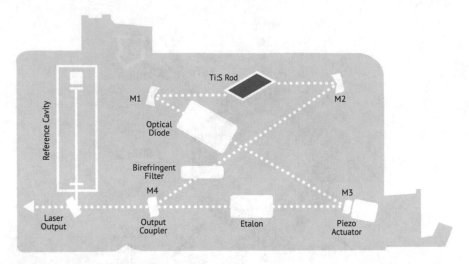

Fig. 4.5 SolsTiS master cavity: Coarse frequency tuning is performed with a birefringent filter (BRF). Applying the etalon lock narrows down the linewidth to <5 MHz. Feedback through the piezo actuator is used for locking to the reference cavity which gives <50 kHz linewidth. Figure from [6]

4.3 Raman Lasers

The Raman lasers are a pair of phase-locked, frequency-doubled, tunable ring resonator CW Ti:Sapphire lasers.[28] They are pumped with a pair of Sprout lasers[29] at 15 W and 10 W. The lasers are tunable over a large range of wavelengths (390 nm $< \lambda <$ 450 nm) but are typically operated at $\lambda \sim 397 - 402$ nm. At these wavelengths the output in the blue is $\sim$2 W for the master laser and $\sim$800 mW for the slave laser.

A sketch of the Ti:Sapph cavity of the master laser is shown in Fig. 4.5, together with the different elements for tuning the frequency. The slave laser does not have its own reference cavity but is instead locked to the master laser via a phase lock. Each Ti:Sapphire output is followed by a pick-off leading to the phase lock photodiode. Part of the picked-off light is led to the wavemeter diagnostic system. The main beams continue to the doubling cavity, and then to the AOM network.

4.3.1 Phase Lock

The phase lock was provided by M squared and has a feedback bandwidth of about 500 kHz. The coarse lock with bandwidth $\lesssim$50 kHz is achieved using the piezo

---

[28]M squared, SolsTiS 1200R + intra-cavity EOM + ECDx 400, and SolsTiS 4000R + ECDx 400.
[29]Lighthouse Photonics, Sprout.

actuator. An additional intra-cavity EOM in the slave laser suppresses noise at higher frequencies. The lasers are phase-locked in the red with a frequency offset of 1.6 GHz, to provide the 3.2 GHz ^{43}Ca$^+$ hyperfine splitting after frequency doubling. The specified mean square phase deviation in the blue integrated for 1 kHz–1 MHz is $\phi^2_{rms} = 7.6 \times 10^{-5}$rad^2.

Alternatively there is a second phase-lock module with the option to measure the beatnote after the frequency doubling cavity, feeding back to a phase offset before the main PID loop. In all experiments in this thesis only the phase lock in the red was used.

4.3.2 AOM Network

The output beam of the Ti:Sapph doubling cavity is fibre coupled into a large mode area fibre[30] so that the AOM network does not need to be realigned after frequency tuning of the Ti:Sapphs or realignment of the doubling cavities. Before the fibre a $\lambda/2$ waveplate followed by a Glan-laser polariser[31] allow directing the majority of the laser power onto a beam dump, for safer alignment. Due to the high laser power cemented optics have to be avoided.[32] The in-coupling face of the fibre for the master laser burned after operation at full power for more than 24 h. It was not replaced but the beam was instead led directly to the AOMs. The fibre for the slave laser did not burn and was used throughout.

We require Raman beams from different directions and at different frequency offsets to perform all necessary tasks (gate beams coupling to motion, sideband cooling beams coupling to motion, motion insensitive carrier drive). All experiments in this thesis were performed using 3 Raman beams: (i) the $R_\parallel$ beam, derived from the master laser, aligned parallel to the magnetic field axis, with $\hat{\sigma}^\pm$ polarisation; (ii) the R_V beam, derived from the master laser, orthogonal to the magnetic field axis, with $\hat{\sigma}^\mp$ polarisation; and (iii) the R_H beam, derived from the slave laser, orthogonal to the magnetic field axis and co-propagating with R_V, with $\hat{\pi}$ polarisation. The beams used for gates are $R_V\&R_\parallel$, for sideband cooling and temperature diagnostics $R_\parallel\&R_H$ and for carrier drive $R_V\&R_H$.

The AOMs for both R_V and $R_\parallel$ are a focussed AOM[33] providing a large frequency scan-range (FWHM = 50 MHz) and a fast rise time (τ_{rise} = 24 ns, 10–90% measured) necessary for fast gates. The R_H AOM[34] is operated at constant frequency. All beams are guided to the trap via polarisation-maintaining single mode fibres[35]

---

[30]NKT Photonics, LMA-PM-5, polarisation maintaining large mode-area single mode fibre.

[31]Thorlabs, GLB10-405, Glan-laser alpha-BBO Polariser.

[32]Friendly tip for the diligent thesis reader: this includes using (cemented) achromatic doublets for fibre coupling...

[33]Brimrose, CQM-200-40-400/OW, crystal quartz, 14 ns nominal risetime.

[34]IntraAction, ASM-2001.588.

[35]Schäfter Kirchhoff, PMC-E-400Si-2.3-NA014-3-APC.EC-200-P.

Table 4.4 Raman beam intensities and waists: The frequencies of the $R_{\parallel}$ AOM vary between different experiments. The beam waists are the $1/e^2$ intensity radii, inferred from measurements of the light shift and carrier Rabi frequencies

Laser	Beam waist (µm)	Polarisation	AOM frequency offset (MHz)
R_V	33	$\hat{\sigma}^{\mp}$	217.368568
$R_{\parallel}$	38	$\hat{\sigma}^{\pm}$	$\sim$220
R_H	29	$\hat{\pi}$	-109

with endcaps so as to be less susceptible to damage caused by the large laser powers. The setup between fibre and trap is similar to that for the other lasers, with a PBS for polarisation clean-up and a pick-off leading to a PD for noise-eating. The R_V and R_H beams are superimposed on a PBS. The $R_{\parallel}$ polarisation is improved with a Glan-Taylor[36] polariser and set with a $\lambda/4$ waveplate on a tip-tilt mount to compensate for birefringence of the viewport window. All Raman beams are focussed onto the trap with $f = 250$ mm lenses.[37] The last mirror of each Raman beam is steered with piezos[38] for more precise alignment. A summary of the Raman beam waists, polarisations and frequency offsets is listed in Table 4.4.

4.3.3   AOM Control

The RF for the AOMs is provided by a DDS board.[39] The setup was inherited from the previous experiment and is described in detail in [3]. In short, each DDS channel has 8 profiles that can be programmed to different frequencies and phase-offsets. The profiles are selected with TTL inputs controlled by the Artiq core device. The DDS output is amplified[40] and mixed[41] with the noise-eater feedback for amplitude stabilisation. The mixer output then goes into a switch[42] that is again controlled by an Artiq TTL pulse to switch the pulse on and off. The output of the switch is amplified[43] to $\lesssim$2.5 W and and then guided to the AOMs. The RF for the R_V-AOM was provided

---

[36]Thorlabs, GT10-A.

[37]Thorlabs, AC254-250-A.

[38]Thorlabs, Polaris-K1S2P.

[39]Enterpoint, Milldown DDS, with Xilinx Spartan 6 LX150T FPGA, and 4 channel AD9910 DDS up to 450 MHz output frequency.

[40]Mini circuits, ZX60-33LN-S+.

[41]Mini circuits, ZP-3MH-S+.

[42]Mini circuits, ZASWA-2-50DR.

[43]Mini circuits, ZHL-03-5WF, high power amplifier, 60–300 MHz, +30 dB gain.

by an AWG[44] for the fast gates, to allow smooth shaping of the amplitude profile with high timing resolution.

4.4 Microwaves and RF

The microwaves are produced with a single-sideband source, explained in detail in [3]. It was slightly modified to accommodate usage at high field. The variable frequency output of a DDS source[45] at f_{DDS} is split with a 90° power splitter.[46] Both arms are fed into an IQ mixer, together with a local oscillator signal at $f_{LO} = 2.77$ GHz produced by a synth.[47] Small DC currents added to the input of the I and Q ports are tuned to improve suppression of the carrier. The desired output is the upper sideband at $f_{\mu w} = f_{DDS} + f_{LO}$. The microwaves are amplified[48] to 10.5 W before they are fed into the vacuum system.

In contrast to micro-fabricated surface traps it is difficult to engineer and precisely predict the field at the ion caused by a microwave antenna in a macroscopic trap. In our system there are therefore 3 possible ways to apply microwaves: an in vacuum antenna pointing towards the trap from the bottom-plate, as well as the two grounded trap blades. Indeed the stretch qubit Rabi frequency for microwaves applied through the antenna was fairly poor for low-field transitions in $^{43}Ca^+$: $t_\pi = 10\,\mu s$ for the low field stretch transition ($\hat\sigma^-$) in $^{43}Ca^+$ ($t_\pi = 100\,\mu s$, 70 μs for $\hat\pi$ and $\hat\sigma^+$ polarised transitions from $|F = 3, m_F = 3\rangle$). Instead the microwaves were applied through one of the grounded blade electrodes, leading to $t_\pi = 3.17(1)\,\mu s$ for the $^{43}Ca^+$ $|F = 4, m_F = 4\rangle \rightarrow |F = 3, m_F = 3\rangle$ transition at 146 G. The microwave polarisation intensity from the ground blade is 28% $\hat\pi$, 30% $\hat\sigma^+$ and 41% $\hat\sigma^-$.

The RF for the strontium Zeeman qubit $5S_{1/2}^{-1/2} \leftrightarrow 5S_{1/2}^{+1/2}$ ($\omega_{0,^{88}Sr^+} = 2\pi \times 408.7$ MHz) is within the reach of a DDS board, and is therefore fed directly from the DDS to an amplifier. It is applied through the second grounded blade.

---

[44] Agilent N8241A, 1.25 GHz clock rate, 15 bits vertical resolution.

[45] Analog Devices, AD9910 evaluation board.

[46] Mini cicruits, ZX10Q-2-3+, 2-way 90° power splitter/combiner.

[47] HP, E4422B, ESG series signal generator, 250 kHz–4 GHz.

[48] Mini circuits, ZHL-16W-43-S+, high power amplifier, 1.8-4 GHz, 45 dB gain.

4.5 Magnetic Field

The main 146 G magnetic field is produced by a pair of water-cooled Helmholtz coils.[49] The power supply for the coils[50] is operated in constant current mode at $I = 56.83$ A. The current noise is measured[51] and stabilised in an active feedback loop [7]. An additional feed-forward loop compensates for ambient magnetic field noise. Two pairs of vertical and horizontal trim coils[52] (see Fig. 4.3) provide additional magnetic fields used to align the B-field axis such that it is superimposed on the 397σ propagation axis. The trim coil currents are $I_{\text{trim,vert}} = 0.18$ A for the vertical trim coils and $I_{\text{trim,horiz}} = 1.8$ A for the horizontal trim coils.

4.6 Experimental Control System

The experiment is controlled using Artiq.[53] At the heart of the control system is the Artiq master, a piece of software that runs experiments, evaluates and saves data, and manages databases and the controller manager. Real-time pulse sequences are compiled and run on the core device, a piece of hardware centred around an FPGA.[54] The core device has TTL outputs that control AOMs, switches, shutters and DDS profiles throughout the lab. They also enable and disable the noise-eater. In addition the core device has a time-resolved TTL input for monitoring APD/PMT counts, with 1 ns timing resolution.

The controller manager handles RPC controllers for different pieces of laboratory equipment, see Fig. 4.1. RPC controllers are used for running the

- piezo controllers[55] for fine-adjustment of the locking cavity lengths, used to scan laser frequencies; also used for control of the piezo-mirrors which centre the Raman beams onto the ions
- trap DAC controller, setting the DC voltages on the trap electrodes and the variable-attenuator adjusting the trap RF amplitude
- tickle synthesizer,[56] setting and scanning the frequency and amplitude of the 'tickle' voltage used to determine trap frequencies
- B-stab controller, running on a beaglebone, controlling the setpoint of the B-field stabilisation as well as the feed-forward parameters

---

[49] Stangeness, rectangular foil-wound coils in custom design, for details see [2].

[50] Keysight, 6671A, 0–220 A output current, rms current noise 200 mA.

[51] LEM, fluxgate sensor, IT 400-S ULTRASTAB.

[52] wire-wound, home-made, vertical: 70 turns, horizontal: 117 turns.

[53] M-Labs, https://m-labs.hk/artiq/Advanced Real-Time Infrastructure for Quantum Physics, DOI: 10.5281/zenodo.591804, version 2.4.

[54] Xilinx, KC705.

[55] Thorlabs, MDT693B, open-loop piezo controller.

[56] Keysight, 33512B waveform generator, 20MHz, 2 channels.

- DDS boards, programming the frequency and phase of the various profiles for the Raman AOM, microwave and RF DDSs
- AWG, programming the pulse amplitude profile for fast gates
- noise eater, setting power set-points for the various laser beams, overwriting the TTL enable/disable and changing the PID parameters for the power stabilisation
- trim coil current supplies[57]
- EMCCD camera, setting the exposure parameters and activating the internal trigger.

The majority of the RPC controllers run on the same main control computer as the Artiq master and communicate to their peripherals via ethernet or USB.

Experimental code is written in a Python-based language, with a restricted set of commands for code run on the core device. Measured fluorescence counts, selected experimental parameters and fitted/calculated results are saved automatically after each successful run of an experiment. The master has a scheduling function that can repeatedly run calibration experiments at fixed time intervals, for example a servo setting the set-point for the magnetic field. Experiments can also be scheduled to run at fixed times, and with different priorities. More details about Artiq can be found in its https://m-labs.hk/artiq/manual-release-3/introduction.htmlonline manual.

References

1. Woodrow SR (2015) Linear Paul trap design for high-fidelity, scalable quantum information processing. Master's thesis, University of Oxford
2. Thirumalai K (2018) Sympathetic cooling and mixed species gates for trapped ion quantum computing. PhD thesis, (in preparation) University of Oxford
3. Ballance CJ (2014) High-fidelity quantum logic in Ca+. PhD thesis, University of Oxford
4. Noursheargh RJ (2016) Miniaturised laser systems for ion trap quantum computing. Master's thesis, University of Oxford
5. Szwer D (2010) High fidelity readout and protection of a 43Ca+ trapped ion qubit. PhD thesis, University of Oxford. papers://d311e016-dabd-41c6-98d5-71ce9eddf36c/Paper/p1856
6. Squared M (2017) Datasheet. http://www.m2lasers.com/images/Datasheet_SolsTiS.pdf
7. Merkel B et al (2019) Magnetic field stabilization system for atomic physics experiments. Rev Sci Instrum 90. ISSN: 044702

---

[57] TTi, QL355TP, power supply.

Chapter 5
Experiment Characterisation

Before entangling gates can be performed on the ions, a number of other operations such as cooling, state preparation, readout and setting the correct magnetic field have to work well. In the following we explain how these operations were implemented and calibrated, and what performance was achieved. We also characterise various bits of equipment and the trapped ion qubits themselves. These measurements are later used to obtain estimates of the error during gate operations.

5.1 Trap Characterisation

5.1.1 Compensation

To minimize the micromotion of the ion, it has to be simultaneously at the centre of both the radial (RF) and the axial (DC) trap. However, machining and assembly imperfections mean that in general these two centres do not perfectly overlap. To move the ion to the ideal spot in the axial direction, asymmetric voltages can be applied to the endcap electrodes. DC voltages applied to the three compensation electrodes can move the ion orthogonally to the trap axis and thus overlap the DC trap minimum with the RF potential null in the radial direction.

First compensation is performed by optimising fluorescence during Doppler cooling. Once the 397 nm fluorescence spectrum looks sufficiently like a half-Lorentzian, the trap can be compensated further by improving RF correlations [1], see Fig. 5.1. This takes advantage of the fact that the Doppler shift caused by the micromotion modulates the ion's fluorescence. The stronger the micromotion, the stronger is the correlation between the detection time of photons emitted by the ion and the phase of the trap RF that gives rise to the micromotion. By minimising this correlation the ion is pushed to a point of weaker micromotion. To minimize micromotion in all three dimensions, a set of three lasers that span all three dimensions has to be

© Springer Nature Switzerland AG 2020
V. M. Schäfer, *Fast Gates and Mixed-Species Entanglement with Trapped Ions*,
Springer Theses, https://doi.org/10.1007/978-3-030-40285-3_5

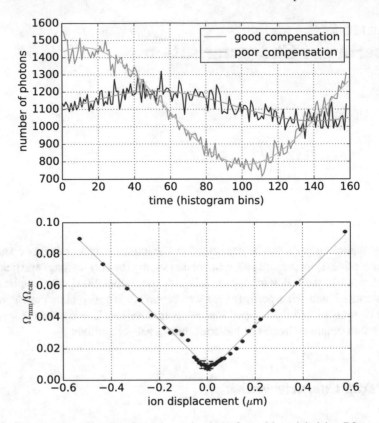

Fig. 5.1 Trap compensation: Rough compensation is performed by minimizing Rf correlations (top). The length of the time bins is given by the resolution of the time-to-analog converter (TAC) and is roughly 0.22 ns. The plot shows the arrival times of fluorescence photons relative to the RF trigger ($t = 0$). The poorly compensated (green) line corresponds to 15 nm micromotion amplitude, whereas the well compensated line (blue) corresponds to roughly 3 nm micromotion amplitude. More precise compensation is performed by minimizing the micromotion excitation relative to the carrier excitation using Raman lasers (bottom). The fitted gradient of the relative axial micromotion amplitude is 0.15(1) per μm

used. The 397 nm laser is detuned to half-fluorescence to get the largest fluorescence modulation for a given Doppler shift. In a final step the endcap voltages are adjusted such that excitation of the ion at the micromotion sideband ($\omega_0 + \omega_{mm}$) is minimal. This can be measured using Raman lasers coupling to the ion's motion. We only optimise this for the axial direction, as all our gates are performed on the axial modes of motion. Due to the gradient of the micromotion along the trap axis only a single ion can sit at the micromotion-null. For two ions centred around the trap centre at the typical trap frequency (ion-ion distance 3.5 μm) the strength of the relative axial micromotion sideband of each ion is $\Omega_{mm}/\Omega_{car} = 0.26$.

Ideal compensation voltages drift over time and have to be updated every few months. Compensation of the weak trap depends strongly on the trap voltages previously applied, indicating thermal effects. For best loading performance the weak trap was therefore always compensated shortly after switching from the tight trap.

5.1.2 Heating Rate

The heating rate of a trap continues to be one of the least well understood mechanisms in ion trapping. While certain relationships are well documented and modelled (e.g. larger ion-electrode distance, lower (cryogenic) trap temperatures and larger (axial) mode frequencies cause a smaller (axial) mode heating rate), the final heating rates of a trap also include a large portion of poorly understood heating, that varies seemingly randomly from trap to trap. This anomalous heating continues to be the subject of investigation [2, 3], with possible candidates being fluctuating patch potentials or lossy dielectric layers.

The axial heating rate at the typical trap frequency ($f_z = 1.92\,\text{MHz}$) in our trap is $\dot{\bar{n}} \approx 75(10)\frac{\text{quanta}}{\text{s}}$ for a single $^{43}\text{Ca}^+$ ion. This is two orders of magnitude larger than the heating rate of the trap's predecessor [4], with similar ion-electrode distances as well as machining and assembly procedures. No noise sources were detected in the voltage supplies for the trap DC. The synth providing the trap Rf was found to be somewhat noisy, but replacing it with a very clean synth did not have any effect on the heating rate.

Heating due to noisy electric fields can be described as [5]

$$\dot{\bar{n}} = \frac{e^2}{4M\hbar\omega_m} S_E(\omega_m) \propto \omega_m^{-\alpha} \tag{5.1}$$

where $S_E(\omega_m)$ is the spectral density of the noise. Its scaling α with ω_m differs between traps. To investigate the large heating rate we have measured α for both the radial and axial modes in our trap, see Fig. 5.2. We obtain $\alpha = 2.8(4)$ for the radial mode heating rate and $\alpha = 2.11(9)$ for the axial mode heating rate. This agrees well with a $1/f$-noise source for the axial heating rate. For the radial heating rate we have neglected the (weaker) heating term due to cross-coupling of the trap RF with the noise fields in the model. The result is therefore still roughly consistent with a $1/f$-type noise source.

The observed increase of the axial heating rate with the RF amplitude is unexpected, because the helical resonator acts as a narrow band-width filter, and noise at the axial mode frequency $f_{\text{noise}} \sim 2\,\text{MHz}$ would be strongly suppressed. That the axial heating rate nonetheless increases with the trap RF suggests there is a mechanism driven by the trap RF causing noise at $\sim 2\,\text{MHz}$. Effects of heating of the trap electrodes or Macor ceramics were excluded by measuring the ion's heating rate directly after a low amplitude trap RF was applied for several hours. The heating rate measured was equal to that when the trap had previously been continuously operated at high RF. Possible suggested reasons explaining the larger heating rate are either

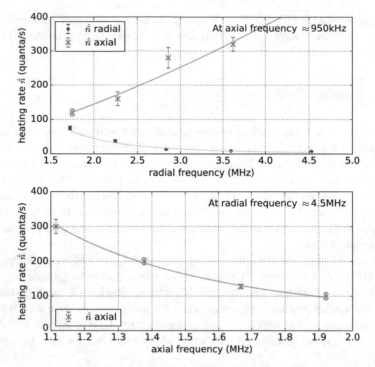

Fig. 5.2 Trap heating rate: The heating rate at the typical trap parameters ($f_z \approx 2$ MHz, $f_{rad} \approx$ 4 MHz) is $\bar{n} \approx 80$ quanta/s and therefore unusually large for a blade trap with 0.5 mm trap-ion distance. As expected the axial (radial) heating rate decreases for larger axial (radial) mode frequency. The axial heating rate also increases for larger radial mode frequency, i.e. larger RF amplitude (with $\alpha = -1.4(2)$)

a partially capacitive solder joint in one of the DC electrode connections, or oxide layers/ patch potentials on the trap electrodes.

The heating rates measured for two ^{43}Ca$^+$ ions are $\dot{\bar{n}}_{ip} = 145(10) \frac{quanta}{s}$ and $\dot{\bar{n}}_{oop} = 0.5(1) \frac{quanta}{s}$. The low heating rate of the oop mode is consistent with a uniform noise source for the observed heating rate.

5.1.3 Ion Order

For mixed species crystals it is important to keep the ion order fixed in all experiments. To set the ion order, whenever a de-crystallisation occurs, the RF amplitude is decreased and a strongly asymmetric endcap voltage is applied. This moves the ions into a radial crystal and away from the trap centre into a less harmonic part of the trap. The RF potential felt by the ions is mass dependent. The axial confinement through the RF therefore tilts the axis of the displaced ion crystal. As the ^{88}Sr$^+$ ion

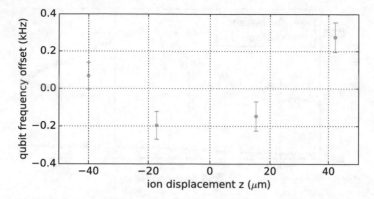

Fig. 5.3 Magnetic field gradient: Change of the qubit frequency of a single ion when it is displaced along the $\hat{z}$-axis of the trap. The ion was moved by applying asymmetric endcap voltages, and the displacement was measured by fitting the ion's position on the camera. The scatter of the measured qubit frequencies is consistent with magnetic field noise, indicating no magnetic field gradient along the trap axis

is heavier it experiences a weaker RF potential and is displaced further from the equilibrium axis, thus leaving the ions in a deterministic order. The RF amplitude is then returned to its normal value and in a last step the endcap voltages are reset. Similar re-ordering techniques were employed for example in [6, 7].

5.1.4 Magnetic Field Gradient

In the predecessor trap of this experiment a large magnetic field gradient of $\frac{df}{dz} = 1.4\,\text{kHz}/\mu\text{m}$ was measured [4], which caused shifts in the qubit frequency for ions at different positions in the trap. The trap materials and machining processes for this trap were chosen carefully to exhibit no magnetic gradient. The qubit frequency was measured over a large range of ion displacements and indeed no magnetic field gradient could be detected, see Fig. 5.3.

5.2 Cooling and Fluorescence

5.2.1 Calcium

The rich level structure of $^{43}\text{Ca}^+$ gives rise to numerous dark resonances that reduce maximum fluorescence and complicate cooling. The large Zeeman splittings at 146 G additionally complicate the conditions, as lasers have to be able to act on population over a larger range of frequencies. By adding sidebands and using separate laser

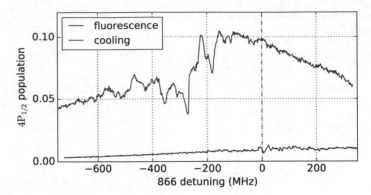

Fig. 5.4 Fluorescence spectrum with dark resonance: The red trace shows the fluorescence of the ion for a scan of the 866 nm laser frequency. At $\delta_{866} = -270$ MHz there is the main dark resonance. At points of large fluorescence the Doppler cooling temperature is fairly high ($\bar{n} \approx 20$). Therefore a second 866_{dark} laser beam, shifted by 270 MHz from the fluorescence laser with an AOM, is used for cooling (blue trace). It is resonant with the dark resonance, when the main 866 nm beam is at a point of high fluorescence. Its intensity is optimised for best cooling and therefore considerably lower than that of the main fluorescence laser

frequencies for cooling and fluorescence detection, sufficiently high fluorescence rates can be achieved nevertheless. Cooling on the dark resonance feature on ^{43}Ca$^+$ can even reach temperatures below the normal Doppler limit.

We follow the dark resonance[1] cooling scheme for ^{43}Ca$^+$ at intermediate field outlined in [8]. However, due to a different trap and beam geometry, we are not bound to a fixed ratio of $\hat{\pi}$ and $\hat{\sigma}^+$ polarisations of the 397 nm laser. Instead we have two separate $397\sigma^+$ and 397π beams, whose relative intensity we can adjust freely. This enables even better cooling performance. In our setup the EOM frequency $f_{\text{EOM,lit}} = 2.930$ GHz used in [8] to bridge the hyperfine splitting between the two ground state levels led to an undesired 'blue-blue' dark resonance. This is a dark resonance caused by the different frequency components of the 397 nm laser only. It manifests itself in slightly reduced fluorescence, poor cooling performance and greatly enlarged state-preparation time constant. To avoid this dark resonance we instead operated the EOM at frequency $f_{\text{EOM}} = 2.955$ GHz.

The laser powers used for fluorescence are $P_{397\sigma^+} = 140\,\mu$W, $P_{397\pi} = 30\,\mu$W and $P_{866} = 2.3$ mW. The resulting fluorescence spectrum can be seen in Fig. 5.4. The peak fluorescence with these parameters is $210{,}000\,\frac{\text{counts}}{s}$, including $12{,}000\,\frac{\text{counts}}{s}$ of background scatter, with a collection efficiency $\eta_{\text{coll}} = 1.5(1)\%$.

---

[1] between the states $4S_{1/2}^{4,+4}$ and $3D_{3/2}^{m_F=5}$ via $4P_{1/2}^{4,+4}$.

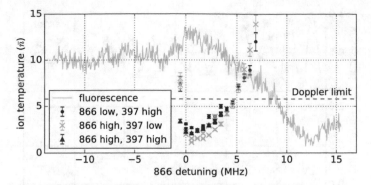

Fig. 5.5 Dark resonance cooling optimisation: Final ion temperature after dark resonance cooling for scanned 866 detuning and different cooling laser powers. The different power settings were (i) $P_{397\sigma} = 9\,\mu\text{W}$, $P_{397\pi} = 5\,\mu\text{W}$, $P_{866d} = 400\,\mu\text{W}$ (dark blue), (ii) $P_{397\sigma} = 4\,\mu\text{W}$, $P_{397\pi} = 2\,\mu\text{W}$, $P_{866d} = 450\,\mu\text{W}$ (light blue), (iii) $P_{397\sigma} = 9\,\mu\text{W}$, $P_{397\pi} = 5\,\mu\text{W}$, $P_{866d} = 450\,\mu\text{W}$ (magenta). The Doppler limit for a $^{40}\text{Ca}^+$ion at the same axial trap frequency $f_z = 1.86\,\text{MHz}$ is shown for comparison. The fluorescence spectrum of the dark resonance for the corresponding 866 detuning is plotted for orientation (with arbitrary units along the y-axis). There is a $\sim$2 MHz frequency uncertainty on the 866 detuning between the fluorescence and temperature scans due to piezo drifts

5.2.1.1 Dark Resonance Cooling

The parameters for dark resonance cooling were optimized empirically, starting from the parameters found in [8]. Iterative scans of the laser powers $P_{397\sigma}$, $P_{397\pi}$, P_{866d} and detunings of the 866 nm and 397 nm lasers, see Fig. 5.5, led to a final temperature of $\bar{n} = 1.25(2)$ after dark resonance Doppler cooling. This corresponds to a temperature of $T = \bar{n}\hbar\omega_z/k_B = 0.11\,\text{mK}$ at the trap frequency $\omega_z = 2\pi \times 1.86\,\text{MHz}$. The ion temperature after Doppler cooling is determined by a simultaneous fit of timescans of the carrier and red- and blue sidebands, see Fig. 5.6. The final powers used for dark resonance cooling are $P_{397\sigma} = 4\,\mu\text{W}$, $P_{397\pi} = 2\,\mu\text{W}$, $P_{866d} = 450\,\mu\text{W}$.

5.2.1.2 Sideband Cooling

For cooling the ion to its ground-state, sideband cooling is employed. We distinguish between continuous and pulsed sideband cooling [9]. For continuous cooling the 397σ, 866, 854 and Raman lasers on the RSB are simultaneously applied to the ion. For pulsed cooling RSB Raman pulses are iterated with repumping pulses of 397σ and 866. The beam powers and pulse lengths were empirically optimised and are $P_{866} = 0.1\,\text{mW}$, $P_{397\sigma} = 1.0\,\mu\text{W}$, with pulse lengths $t_{\text{RSB}} = 12\,\mu\text{s}$ and $t_{\text{repump}} = 60\,\mu\text{s}$. At the final temperature $t_{\pi,\text{RSB,final}} \approx 18\,\mu\text{s}$. The decrease in ion temperature after different types of cooling is shown in Fig. 5.7.

The ion temperature after sideband cooling is estimated from the relative heights of the red- and blue sideband with [9, 10]

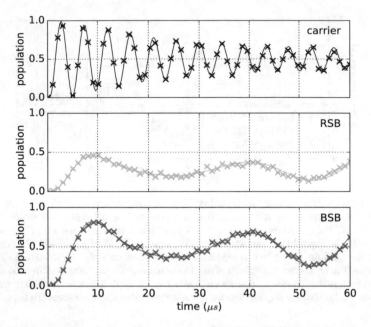

Fig. 5.6 Dark resonance cooled ion: Time scan of carrier, RSB and BSB flops of a dark-resonance Doppler cooled ion. The solid lines are a simultaneous fit of all three curves, resulting in $\bar{n} = 1.25(2)$, with other floated parameters $t_\pi = 2.252(2)\,\mu$s, $t_{\text{dead}} = 0.37(2)\,\mu$s and $\eta = 0.1786(3)$. The final temperature at this point is very sensitive to drifts of the 866 nm laser frequency—already small drifts can cause Doppler-heating of the ion. The 866 nm laser frequency is therefore usually set slightly higher to be more robust to drifts. The final Doppler cooling temperature is then typically $\bar{n} \approx 1.8$

Fig. 5.7 Ground state cooling of a single ion: The ion is initially cooled with dark resonance cooling for $t_{\text{DRC}} = 0.6$ ms to $\bar{n} = 1.79(7)$ (top). It is then cooled further with continuous sideband cooling for $t_{\text{CSBC}} = 1.2$ ms to $\bar{n} = 0.027(4)$ (bottom). Pulsed sideband cooling did not show any significant advantage. The fitted cooling time constants are $\tau_{\text{DRC}} = 60(20)\,\mu$s and $\tau_{\text{CSBC}} = 150(10)\,\mu$s

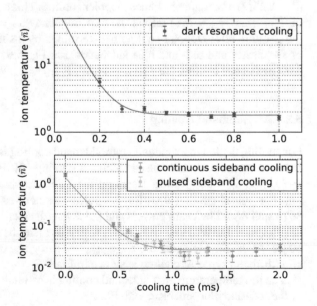

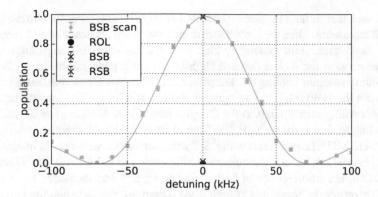

Fig. 5.8 Sideband cooled ion: Temperature diagnostics for a sideband cooled ion is performed by measuring the maximum inversion of a BSB and a RSB π-pulse, compared to the readout level (ROL). The light blue line shows a frequency scan over the BSB of a single sideband cooled ion at $\bar{n} = 0.017(3)$

$$\bar{n} = \frac{r}{1-r} \tag{5.2}$$

where the ratio $r = p_{\mathrm{RSB}}/p_{\mathrm{BSB}}$. This measurement is insensitive to slight errors in the pulse length or frequency of the pulses, as long as they are identical for both sidebands. We therefore measure only the top of the RSB and BSB, as shown in Fig. 5.8. For two ions we again measure only the peak inversion population of the RSB and BSB. The ion temperature $\bar{n}$ is then fitted, such that the measured populations agree with those predicted for a thermal state.

For sideband cooling two ions the two different (ip and oop) modes have to be cooled separately. To reduce effects of heating, we first cool the oop and then the ip mode. A cooling sequence then looks like $\mathrm{DRC}(t_{\mathrm{drc}}) - \mathrm{CSBC}_{\mathrm{oop}}(t_{\mathrm{csbc}}) - \mathrm{CSBC}_{\mathrm{ip}}(t_{\mathrm{csbc}}) - \mathrm{PSBC}_{\mathrm{oop}}(n_{\mathrm{psbc}}) - \mathrm{PSBC}_{\mathrm{ip}}(n_{\mathrm{psbc}})$ with $t_{\mathrm{drc}} = 1\,\mathrm{ms}$, $t_{\mathrm{csbc}} = 1\,\mathrm{ms}$ and $n_{\mathrm{psbc}} = 10$ for two ions.

During the optimisation of fast gates the gate fidelity appeared to be highly sensitive to the 866 detuning, even though the final ion temperature seemed unaffected measured with the normal temperature diagnostic. Increasing the cooling length to $t_{\mathrm{drc}} = 2.5\,\mathrm{ms}$, followed by $t_{\mathrm{sbc}} = 1\,\mathrm{ms}$ and $n_{\mathrm{psbc}} = 10$ removed the strong dependency on the 866 detuning. We believe the high sensitivity of the fast gate experiments is due to the large displacements in phase space and enhancement of out-of-Lamb-Dicke effects for small parts of the population in high-n states.

5.2.2 Strontium

The level structure in $^{88}\mathrm{Sr}^+$ is significantly simpler, allowing the usual Doppler cooling and fluorescence routines similar to $^{40}\mathrm{Ca}^+$ (see e.g. [9]). The 1092 detuning is chosen such that its dark resonance with the 422 laser is tuned slightly into the blue of the 422 resonance. Thus the ion's fluorescence spectrum for scanned 422 frequency is half-Lorentzian, with maximally high peak fluorescence. For high fluorescence Doppler cooling the 422 is detuned slightly red from peak fluorescence. For low temperature Doppler cooling the 422 power is reduced such that this detuning coincides with the HWHM, yielding optimal cooling efficiency. The temperature after Doppler cooling corresponds to the Doppler limit $T_D = \hbar\Gamma/(2k_B) \approx 0.5\,\mathrm{mK}$, corresponding to $\bar{n} \approx 8$ for a single $^{88}\mathrm{Sr}^+$ ion at its typical axial frequency $\omega_{z,^{88}\mathrm{Sr}^+} = 2\pi \cdot 1.3413\,\mathrm{MHz}$. Experiments with $^{88}\mathrm{Sr}^+$ were performed with the old imaging system in place which had a lower collection efficiency $\eta = 0.282(7)\%$ (at 397 nm) [4]. In addition, the fluorescence of both $^{43}\mathrm{Ca}^+$ and $^{88}\mathrm{Sr}^+$ was measured on the same PMT. Therefore the focus was chosen to be a compromise between the two different wavelengths, slightly reducing the signal to noise ratio. The fluorescence rate measured from $^{88}\mathrm{Sr}^+$ was $33,300\,\frac{\mathrm{counts}}{s}$, including $4,600\,\frac{\mathrm{counts}}{s}$ background.

5.2.3 Recrystallisation

Collisions with background gas atoms in the vacuum chamber can cause the ions to heat up or even de-crystallise. To confirm that the ions are still cold and crystallised, at the end of each readout event, ion fluorescence is collected with the 854 nm deshelving laser enabled. If the collected fluorescence is below the threshold for a bright ion the last data point is discarded and the ions are Doppler cooled. If this happens three times in succession it is assumed that the ions have de-crystallised and an automatic recrystallisation is performed. For this the trap is set to the weak trap for 1.5 s with Doppler cooling enabled. Afterwards the trap is switched back to the standard tight trap.

5.3 State Preparation and Readout

5.3.1 Calcium

As described in Sect. 2.3.1 state preparation in $^{43}\mathrm{Ca}^+$ is performed with polarisation-selective optical pumping. The state-preparation fidelity is therefore defined by the polarisation impurity of the $397\sigma^+$ light. To measure the polarisation impurity, the ion is prepared in $|\!\downarrow\rangle$. An intense 397σ pulse is applied on the ion, followed by shelving using only the 393 laser. Polarisation impurities in the 397σ pulse transfer population into $3D_{3/2}$ and appear bright during readout. The polarisation purity is optimised through minimizing this bright signal by adjusting a $\lambda/4$ wave-plate on a tip-tilt mount and tuning the current through the trim-coils to align the magnetic field axis exactly parallel to the 397σ beam.

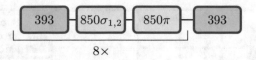

Fig. 5.9 Shelving sequence: Population is transferred to $3D_{5/2}$ via $4P_{3/2}$, $m_F = 5$ with a 393 nm pulse. Population that has decayed to $3D_{3/2}$ is recovered with two 850 nm pulses. This sequence is repeated 8 times for optimum readout fidelity. The pulse lengths are $t_{393} = 12\,\mu s$, $t_{850\sigma_{1,2}} = 6\,\mu s$ and $t_{850\pi} = 2\,\mu s$

For shelving we employ the pulsed sequence described in Fig. 5.9 and developed in [11]. The state of the ion is then determined by applying fluorescence lasers, and counting emitted photons for a duration t_{bin}. If the number of detected photons is below the threshold, the ion is assumed to be 'dark', i.e. in the shelf. Otherwise it is counted as 'bright'. The threshold is calculated with

$$\text{threshold} = \frac{\bar{c}_b}{\ln(1 + \bar{c}_b/\bar{c}_d)} \qquad (5.3)$$

where $\bar{c}_b$ ($\bar{c}_d$) are the mean bright (dark) counts measured over a large set of repetitions. For the bright counts fluorescence of an un-shelved Doppler cooled ion is measured with all re-pumping lasers switched on. For the dark counts the 866 nm repumping laser is switched off. For two ions the threshold between 1 ion bright ($p_{\downarrow\uparrow} + p_{\uparrow\downarrow}$) and two ions bright ($p_{\uparrow\uparrow}$) is set to be in the centre of the two mean fluorescence counts, see Fig. 5.10.

The readout fidelity is limited by the lifetime of the shelf as well as polarisation impurities or off-set frequency of the shelving lasers. The 393 nm laser is superimposed with the 397σ laser, and its polarisation is set by optimising the 397σ polarisation. The 850σ polarisation is optimised with a $\lambda/4$ waveplate on a tip-tilt mount. The polarisation impurity is measured by using the 850σ as repumper from $3D_{3/2}$ together with the 854 nm laser and then minimising fluorescence. For perfect polarisation, population will be pumped into $3D_{3/2}$, $m_F = 5$ and the ion will go dark. The 850π polarisation is set by geometry, by using a linearly polarised beam orthogonal to the magnetic field axis.

The frequency of the 393 nm laser is set by preparing the ion in $|\downarrow\rangle$ and shelving it with a single pulse of 393 nm light of length $t_{393} = 12\,\mu s$. The frequency of the 393 nm laser is scanned and then set to the centre of the fitted Lorentzian. The 850 nm laser frequency is set using the 850π. First the ion is prepared into $|\downarrow\rangle$, and then into $3D_{3/2}$, $m_F = 5$ using a long 1.5 ms pulse of 393, 397σ, 854 and $850\sigma_{1,2}$ light. Subsequently a short $2\,\mu s$ 850π pulse is applied with scanned frequency. If on resonance, it transfers population back into the ground-state manifold. From there it is transferred into $3D_{5/2}$ by a 393 shelving pulse and will appear dark in a fluorescence measurement. The curve is again fitted with a Lorentzian and the 850π set to its centre frequency.

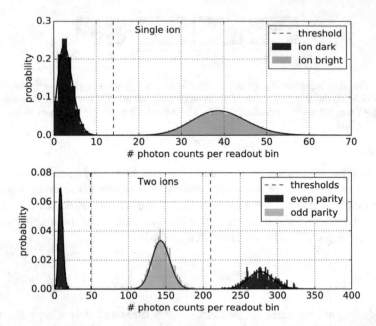

Fig. 5.10 Readout histograms: Top: Histograms for a single ion, with normalised probabilities. The histograms of each ion state consist of 50.000 data points. The readout bin length is $t_{\mathrm{bin,1ion}} = 230\,\mu$s. Bottom: Readout histograms for two ions of even ($|\uparrow\uparrow\rangle + |\downarrow\downarrow\rangle$) and odd parity ($|\uparrow\downarrow\rangle + |\downarrow\uparrow\rangle$) Bell states. The readout bin length is $t_{\mathrm{bin,2ions}} = 800\,\mu$s. It is longer than the single ion bin length to obtain a large enough separation between the '1-ion-bright' (odd parity) and '2-ions-bright' (even parity, right) histograms

The fidelities we achieved in combined state-preparation and readout measurements (SPAM) for a single $^{43}\mathrm{Ca}^+$ ion are $\varepsilon_{\mathrm{LRL}} = 6(1) \times 10^{-4}$ for $|\downarrow\rangle = 4S_{1/2}^{4,+4}$ and $\varepsilon_{\mathrm{URL}} = 1.0(2) \times 10^{-3}$ for $|\uparrow\rangle = 4S_{1/2}^{3,+3}$, leading to a total SPAM error of $\varepsilon_{\mathrm{SPAM}} = \frac{1}{2}(\varepsilon_{\mathrm{LRL}} + \varepsilon_{\mathrm{URL}}) = 8(1) \times 10^{-4}$.

5.3.2 Strontium

Experiments performed with strontium were only first proof-of-principle experiments, with only functional readout. State preparation is performed with frequency selective pumping with the $422l$ ($422u$) beam at its low power setpoint, with $t_{\mathrm{sp}} = 100\,\mu$s. The simulated error for state-preparation is $\epsilon_{\mathrm{sp}} = 5 \times 10^{-4}$. Because the 674 nm laser was not yet available, a preliminary low-fidelity readout scheme was used, employing a 408 nm laser for shelving. For this the 408 and 1092 lasers are applied simultaneously for $t_{\mathrm{shelve,}^{88}\mathrm{Sr}^+} = 16\,\mu$s. The scheme is limited by off-resonant excitation out of $|1\rangle = 5S_{1/2}^{-1/2}$ and population lost after repumping from $4D_{3/2}$.

The simulated error of the scheme with these parameters is $\epsilon_{\text{ro,sim}} \approx 7.5 \times 10^{-2}$. In the experiment we achieved a combined state preparation and readout error of $\epsilon_{\text{SPAM,exp}} = 10(1) \times 10^{-2}$, strongly dominated by the preliminary readout.

5.4 Magnetic Field

Magnetic field stability is an important requirement for trapped ion experiments, as magnetic field noise is in general the largest source of qubit decoherence. While clock transitions are far less sensitive to magnetic field noise, preparation into the clock qubit still uses magnetically sensitive transitions. In addition, for work with different ion species, the B field will only be at a point of magnetic insensitivity for one ion species. Because all work in this thesis was performed on the magnetically sensitive 'stretch'-qubit, magnetic field stability has a particularly high priority. We care both about the long-term stability of the absolute magnetic field perceived by the ion, as well as fast fluctuations. Slow drifts and changes of the magnetic field are the result of mostly external causes, such as elevators in the building moving or other research groups in the vicinity ramping their poorly isolated magnets. Fast fluctuations are caused by noise in the current generating the field and ambient electromagnetic radiation broadcast by other laboratory equipment. Slow drifts are compensated by a field servo, that adjusts the coil current setpoint such that the ion's qubit frequency remains at a constant value. Current and ambient noise are reduced with a stabilisation circuit characterised in [12].

5.4.1 B-Field Servo

For servoing the magnetic field a microwave pulse of low intensity is used to probe the 'stretch'-qubit transition $4S_{1/2}^{4,+4}$-$4S_{1/2}^{3,+3}$ of a $^{43}\text{Ca}^+$ ion. The frequency of the pulse is set to $f_{\mu w} = f_{\text{stretch}} \pm \Delta f = f_{\text{stretch}} \pm 0.399343/t_\pi$, such that for correct magnetic field the resonance is probed at the two FWHM points. If the magnetic field has drifted off, the populations at the two points are asymmetric. From the population difference and the gradient $m = 1.89709(p_{\text{max}} - p_{\text{min}})t_\pi$ we can calculate the frequency offset.

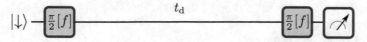

Fig. 5.11 Ramsey sequence with scanned frequency: For precisely measuring transition frequencies, a Ramsey sequence is used where the frequency of both $\pi/2$-pulses is scanned over resonance. A longer Ramsey delay t_d increases the frequency resolution by reducing the width of the fringes. Detuning from the target frequency is inferred from the offset of the centre-fringe from expected resonance

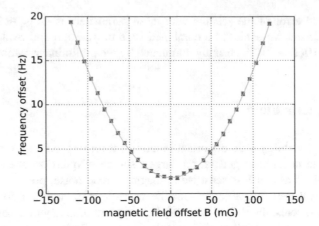

Fig. 5.12 Clock qubit: Transition frequency of $|F = 4, m_F = 0\rangle \leftrightarrow |F = 3, m_F = 1\rangle$ around the magnetic field insensitive point at $B \approx 146$ G. The 0 points of the magnetic field and the frequency are the calculated values from the Breit-Rabi formula. The $\delta f = 1.8(1)$ Hz offset of the clock frequency is due to imperfect calibration of the reference clock of the microwave source and AC Zeeman shifts caused by the trap RF [11]

The setpoint of the field coil current is then adjusted to cancel the measured offset. This measurement is repeated until both measured populations are within 2 standard deviations of $p = 1/2$. If the magnetic field has drifted far enough that neither probe pulse is close to resonance with the transition, a coarse servo is performed first with a shorter t_π. We use $t_{\pi,\mathrm{ls}} = 4.34\,\mu$s for low sensitivity and $t_{\pi,\mathrm{hs}} = 64.8\,\mu$s for high sensitivity. With $N = 300$ repeated measurements on each side of the peak, this sets the magnetic field to within ± 0.2 mG accuracy (± 500 Hz on the stretch transition).

The setpoint of the magnetic field is chosen such that the transition $|F = 4, m_F = 0\rangle \leftrightarrow |F = 3, m_F = 1\rangle$ is at its field-insensitive point (Fig. 5.12). This was confirmed by measuring the clock transition frequency as a function of the B-field. The field was scanned by adding an offset to the target frequency of the stretch qubit in the magnetic field servo. The clock qubit transition frequency was measured using a Ramsey sequence, see Fig. 5.11, with $t_\mathrm{d} = 25$ ms and $t_{\pi/2} = 686\,\mu$s. Choosing a longer $t_{\pi/2}$ reduces the total bandwidth of the Ramsey sequence and therefore simplifies identification of the centre fringe. Methods for identifying the centre fringe and choosing sensible pulse and delay lengths are described in detail in [11].

5.4.2 Field Stability

The current applied to the magnetic field coils is stabilised with an active feedback loop. An additional feedforward circuit parallel to the coils compensates for ambient noise. The stabilisation circuit was designed and built by Benjamin Merkel and is

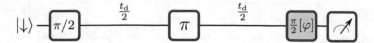

Fig. 5.13 Spin-echo sequence with scanned phase: Coherence times are measured with the above spin-echo sequence (or Ramsey sequence by dropping the π-pulse in the centre). The Ramsey contrast is determined by scanning the phase of the final $\pi/2$-pulse $\varphi : 0 \mapsto 2\pi$ and fitting the resulting sine-function

described in detail in [12]. It reduces field noise at $146\,\text{G}$ to $45(1)\,\mu\text{G}$ rms noise, corresponding to $0.31(1)$ ppm.

We can measure the resulting coherence time of the qubit with a Ramsey phase-scan experiment, see Fig. 5.13. Fluctuations of the qubit frequency during the Ramsey delay t_d cause dephasing and lead to a reduction of contrast after the final $\pi/2$ pulse. The π pulse in the centre of the spin-echo sequence reverses the dynamics, and therefore leads to a cancellation of fluctuations slow relative to the delay t_d. The spectral density of the noise affects the pattern of the contrast decay [13]. White noise or noise that is correlated over durations much shorter than the Ramsey delay leads to exponential decay of the contrast $c = a \cdot e^{-t_d/\tau}$. Noise with correlation times long compared to the Ramsey delay leads to a Gaussian contrast decay $c = a \cdot e^{-(t_d/\tau)^2}$. Measurements of the coherence time both in Ramsey and spin-echo geometry are shown in Fig. 5.14.

5.5 Raman Lasers

Raman lasers are used for cooling as well as two-qubit operations. Their stable and well controlled operation is therefore vital for (high-fidelity) ion manipulation. The effect of the Raman light can be fully described by its frequency, phase, polarisation and intensity and wave-vector.

Frequency Due to the large Raman detuning ($\Delta = 200\,\text{GHz}\text{--}10\,\text{THz}$) typical frequency fluctuations have negligible effects on the ion and the Raman laser frequency does not have to be controlled carefully. Relevant for gates is only the relative frequency difference between the two Raman lasers, which is set with AOMs and can therefore be controlled very precisely.

Phase For two-qubit gate operations both Raman beams are derived from a single laser, and relative phase fluctuations are only caused by path-length fluctuations of the different beams. These have acoustic sources and are therefore slow compared to a single gate pulse. They therefore only affect gate sequences on a shot-to-shot timescale. We do not stabilise these slow phase fluctuations in our experiment, leading to a random initial Raman phase ϕ_0. The AOMs used for switching can also cause phase chirps. This effect is discussed in Sect. 7.3.2. For single-qubit operations beams

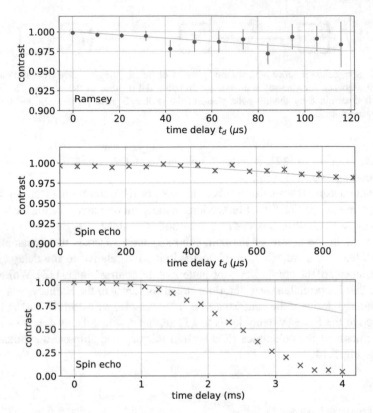

Fig. 5.14 Spin coherence of the stretch state qubit, measured in a Ramsey (top) and a spin-echo sequence (centre). The fitted coherence time for small errors is $\tau_{\text{Ramsey}} = 4.8(9)$ ms for the Ramsey sequence, fitted to exponential decay with fixed $a = 1$. Low-frequency drifts increase the errors for larger Ramsey delays, as can be seen in the larger error bars. For the spin-echo sequence the noise-spectrum is filtered and a Gaussian decay fits better to the data. This fits to $\tau_{\text{spin}-\text{echo}} = 6.3(3)$ ms, with floated scale-factor $a = 0.9984(8)$. For larger delays other noise mechanisms lead to a faster decay of the fringe contrast (bottom). The coherence times fitted with the above methods assume a simplified noise-spectrum and vary slightly with the maximum delay time. However, because for our gates the coherence behaviour at small times ($t_d \lesssim t_g$) is most relevant, they provide an upper bound for errors due to spin-decoherence

from the two different Raman lasers are necessary. They rely on the phase stability between both lasers assured by their phase lock.

Polarisation Setting the polarisation of the Raman lasers accurately is important for performing maximally efficient two-qubit gates and minimising the single-beam light shifts. The R_V and R_H beams are linearly polarised orthogonally to the magnetic field axis. Since only an asymmetry of $\hat{\sigma}^+$ and $\hat{\sigma}^-$ polarisation will cause a differential light shift (see Appendix B), this beam geometry ensures that imperfections in the polarisation of these two beams won't cause a differential light shift. Their linear polarisation is therefore simply set by a PBS. The $R_{\parallel}$ beam however is propagating

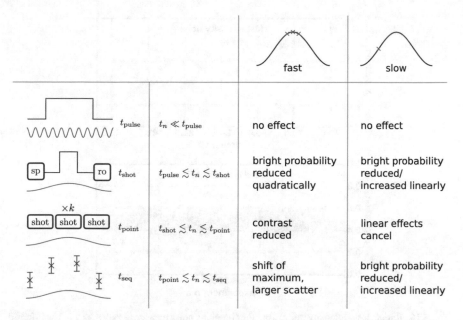

		fast	slow
t_{pulse} — $t_n \ll t_{\text{pulse}}$		no effect	no effect
t_{shot} — $t_{\text{pulse}} \lesssim t_n \lesssim t_{\text{shot}}$		bright probability reduced quadratically	bright probability reduced/ increased linearly
t_{point} — $t_{\text{shot}} \lesssim t_n \lesssim t_{\text{point}}$		contrast reduced	linear effects cancel
t_{seq} — $t_{\text{point}} \lesssim t_n \lesssim t_{\text{seq}}$		shift of maximum, larger scatter	bright probability reduced/ increased linearly

Fig. 5.15 Intensity noise measurement: The above table shows the different timescales in the slow and fast intensity noise measurement, and how noise of different frequencies (indicated by the red sine-waves of period $\sim t_n$) affects the measurement outcome. In both measurements a single shot consists of cooling and state-preparation (sp), followed by a Raman pulse of length t_{pulse} and readout (ro). One measurement point consists of k identically performed shots. For measurements of fast noise $\Omega_R t_{\text{pulse}} = (2n + 1)\pi$ and k is large to obtain a large average over populations. For one sequence, t_{pulse} is scanned over the top of a fringe and the maximum contrast is measured. For measuring slower noise $\Omega_R t_{\text{pulse}} = 2\pi n + \pi/2$ and k is smaller to increase the range of noise frequencies the experiment is sensitive to. In one sequence each point is repeated a large number of times with constant t_{pulse}. The noise can be inferred from the scatter of the populations

parallel to the magnetic field axis. Slight imperfections in the beam's polarisation can therefore easily create an asymmetry of $\hat{\sigma}^+$ and $\hat{\sigma}^-$ polarisation and give rise to a large light shift. Consequently the polarisation of the $R_\parallel$ is set more carefully, using a Glan-laser polariser, followed by a $\lambda/4$ wave-plate on a tip-tilt mount to compensate for birefringence in the viewport window. The polarisation is optimised by applying an $R_\parallel$-pulse within a Ramsey interferometer and minimising the phase acquired due to the light shift. The achieved polarisation purity is such that the residual light shift corresponds to $\lesssim 4 \times 10^{-5}$ of the maximum possible lightshift.

Intensity The intensity at the ion depends on the Raman beam power as well as beam pointing onto the ions. We measure intensity noise on two different timescales f_n, see Fig. 5.15: fast noise on the timescale of single Raman pulses reduces the contrast of Rabi oscillations. Determining the maximum contrast after a large number of fringes (n) therefore gives a sensitive measure of fast intensity noise. The time scales in our experiment were $t_{\text{pulse}} \approx 90\,\mu\text{s}$, $t_{\text{shot}} \approx 4\,\text{ms}$, $t_{\text{point}} \approx 8\,\text{s}$ and $t_{\text{seq}} \approx 80\,\text{s}$ with $k = 2000$. This sequence is therefore sensitive for noise frequencies $100\,\text{mHz} <$

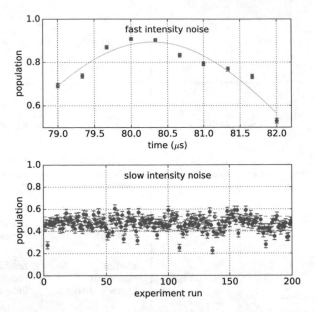

Fig. 5.16 Raman laser intensity noise: Top: Fast intensity noise measured with $n = 11$. The inferred relative intensity noise is $\delta\Omega/\Omega_R = 8.6(9) \times 10^{-3}$ for a well-aligned beam. For a beam aligned for maximum sensitivity to beam-pointing noise $\delta\Omega/\Omega_R = 2.3(4) \times 10^{-2}$, measured at the 5th fringe ($n = 5$). Bottom: Slow intensity noise measured on the 6th fringe for a well aligned beam. The relative noise is $\delta\Omega/\Omega_R = 3.2 \times 10^{-3}$ inferred from the standard deviation of the measured points (red dotted lines). For maximum beam-pointing sensitivity we obtain on the 3rd fringe $\delta\Omega/\Omega_R = 1.3 \times 10^{-2}$

$f_n < 11$ kHz. Slower noise can be measured more accurately by looking at the scatter of points in the linear range of a fringe. Here we use $t_{\text{shot}} \approx 4$ ms, $t_{\text{point}} \approx 0.8$ s, $t_{\text{seq}} \approx 160$ s with $k = 200$. This sequence is therefore sensitive to noise with 6 mHz $< f_n < 1.2$ Hz. Results of the noise measurements are shown in Fig. 5.16.

As in gate sequences, the Raman lasers are switched on for $t = 100\,\mu$s with noise-eating enabled at the beginning of each state preparation block. This stabilises the Raman laser power measured from a beam pick-off shortly before the trap. Between the power stabilisation and the Raman pulse used to measure the intensity noise, there is a ~ 2 ms delay during which the ion is cooled and during which no further power stabilisation is active. Raman power noise in timescales of $t_n \gg t_{\text{shot}}$ is therefore taken out by the noise-eater. The sequence does however measure intensity noise due to beam pointing fluctuations at these timescales, which cannot be detected by the noise eater.

Effects of beam pointing noise are enhanced for a slightly mis-aligned beam, where the intensity gradient of the Gaussian beam-profile is larger. For best beam alignment performance we therefore use piezo mirrors to peak up the beam. A Raman pulse of $t = (2n + \frac{1}{2})\pi$ is shone onto the ion and the voltage applied to a piezo mirror is scanned. The setpoint voltage is adjusted to maximise the measured population until there is a turning point at $p = 0.5$ at the voltage setpoint. The experiment is repeated up to $t = 10.5\pi$, which corresponds to alignment of the Raman beam to within $0.04r_0$, where r_0 is the beam waist at the ion.

The two-qubit gate error scales quadratically with the relative Rabi-frequency uncertainty, leading to $\varepsilon_g \approx 2 \times 10^{-4}$ for the intensity noise measured here. For faster gates this can deviate, as different timescales of noise are relevant.

5.6 Motional Coupling to Raman Lasers

5.6.1 Ion Spacing

The phase difference of the light-shift force on two ions depends on their spacing relative to the wavelength of the 'travelling standing wave' of the light, see Sect. 3.4.2. For an optimally efficient gate, only states with either even parity spins ($|\uparrow\uparrow\rangle$, $|\downarrow\downarrow\rangle$) or only states with odd parity spins ($|\uparrow\downarrow\rangle$, $|\downarrow\uparrow\rangle$) should be excited. This is the case if the ion spacing matches $n\lambda_z$ or $(n + \frac{1}{2})\lambda_z$, where λ_z is the periodicity of the standing wave and n a natural number. If the ion spacing is between these values, all four spin states are excited and acquire a geometric phase. This would mean the geometric phase partially translates into a global phase, rather than only into the desired two-qubit phase. To prevent this, we carefully set the ion spacing to match a half-integer standing wave length. The ion spacing is optimised by preparing the ions in $|\downarrow\downarrow\rangle$ and exciting their motion with a light-shift force either resonant with the ip or the oop axial mode. The resulting excitation is then measured by probing the red and blue sideband excitation, see Fig. 5.17. The endcap voltages are adjusted to minimise (maximise) excitation of the ip (oop) mode, corresponding to half-integer spacing. The absolute ion spacing can be inferred from the axial trap frequency with Eq. 3.6 and is $d_{\text{ion-ion}} = 12.5\lambda_z$.

5.6.2 Motional Coherence

During a gate operation the ions are not only in a superposition of spin states, but also in a superposition of motional states. It is therefore important to determine the dephasing of motional states to calculate the ensuing errors. Following [14], we measure motional coherence with a modified Ramsey sequence, see Fig. 5.18.

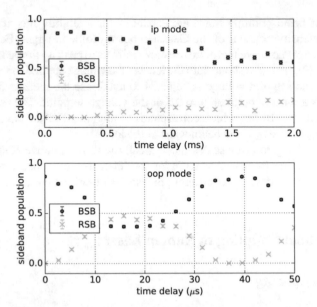

Fig. 5.17 Ion spacing: Motional excitation of the ip (top) and oop (bottom) mode for a $12.5\lambda_z$ ion spacing. For half-integer ion spacing the force felt by two equal spins is opposite. This leads to strong excitation of the oop mode, but negligible excitation of the ip mode. The parasitical excitation of the ip mode is less than a percent of the oop mode excitation. Additional scattering errors, that would be caused by the reduced gate efficiency due to parasitical mode excitation, are therefore negligible [4]

Fig. 5.18 Motional coherence sequence: The motional coherence is measured with a Ramsey (spin-echo) sequence where the spin-state superposition is mapped onto a superposition of motional states of the same spin-state using red sideband π-pulses π_R. For a Ramsey measurement the centre block is removed

Motional frequency noise during the delay leads to decay of the Ramsey contrast, as can be seen in Fig. 5.19. The fitted coherence time is $\tau_{mot} = 8.8(9)$ ms, and does not differ between a Ramsey and spin-echo measurement. This indicates that motional dephasing is not dominated by slow drifts, but by white noise or noise with short correlation times. Indeed, for small errors, simulation of motional dephasing due to the trap's heating rate shows a $\tau_{mot} \approx 1/(2\dot{\bar{n}}) \approx 7$ ms dependence. This suggests that the motional coherence in our experiment is limited by the heating rate of the trap, both in Ramsey and spin-echo geometry.

Fig. 5.19 Motional coherence: Measurement of the motional coherence of a single ion, in a Ramsey and spin-echo interferometer. The motional coherence is equal for both measurements, because it is dominated by motional heating which is not reversed through a spin-echo geometry

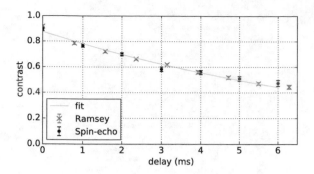

References

1. Berkeland DJ, Miller JD, Bergquist JC, Itano WM, Wineland DJ (1998) Minimization of ion micromotion in a Paul trap. J Appl Phys 83:5025–5033. ISSN: 00218979
2. Brownnutt M, Kumph M, Rabl P, Blatt R (2015) Ion-trap measurements of electric-field noise near surfaces. Rev Mod Phys 87:1419–1482. ISSN: 15390756
3. Sedlacek JA et al (2018) Distance scaling of electric-field noise in a surface electrode ion trap. Phys Rev A 97:020302. ISSN: 24699934
4. Ballance CJ (2014) High-fidelity quantum logic in Ca+. PhD thesis, University of Oxford
5. Turchette QA et al (2000) Heating of trapped ions from the quantum ground state. Phys Rev A 61:063418. ISSN: 1050-2947
6. Hume DB, Chou CW, Rosenband T, Wineland DJ (2009) Preparation of Dicke states in an ion chain. Phys Rev A 80:052302. ISSN: 10502947
7. Home JP, Hanneke D, Jost JD, Leibfried D, Wineland DJ (2011) Normal modes of trapped ions in the presence of anharmonic trap potentials. New J Phys 13:073026. ISSN: 13672630
8. Allcock DTC et al (2016) Dark-resonance Doppler cooling and high fluorescence in trapped Ca-43 ions at intermediate magnetic field. New J Phys 18
9. Webster S (2005) Raman sideband cooling and coherent manipulation of trapped ions. PhD thesis, University of Oxford
10. Monroe C et al (1995) Resolved-sideband Raman cooling of a bound atom to the 3D zero-point energy. Phys Rev Lett 75:4011–4014. ISSN: 00319007
11. Harty TP (2013) High-fidelity microwave-driven quantum logic in intermediate field 43Ca+. PhD thesis, University of Oxford
12. Merkel B et al (2019) Magnetic field stabilization system for atomic physics experiments. Rev Sci Instrum 90. ISSN: 044702
13. O'Malley PJJ et al (2015) Qubit metrology of ultralow phase noise using randomized benchmarking. Phys Rev Appl 3:044009. ISSN: 23317019
14. Turchette QA et al (2000) Decoherence and decay of motional quantum states of a trapped atom coupled to engineered reservoirs. Phys Rev A 62:053807. ISSN: 10502947

Chapter 6
Mixed Species Gates

Entangling ions of different species is an important prerequisite for transferring quantum information between them. Thus we can choose to perform operations on the best suited species, harnessing their individual strengths. We use the σ_z geometric phase gate that only requires a single pair of Raman beams to perform a gate simultaneously on two different species. In a first test of the scheme we perform the gate between two different isotopes of calcium: ^{40}Ca$^+$ and ^{43}Ca$^+$. These results were published in [1] and are also discussed in parts in [2]. In a proof-of-principle experiment we then realise the same gate on two different atomic species—^{43}Ca$^+$ and ^{88}Sr$^+$.

6.1 ^{40}Ca$^+$–^{43}Ca$^+$ Gate Results

The experiments with ^{43}Ca$^+$ and ^{40}Ca$^+$ were performed in the predecessor trap to the one described and characterised in this thesis. The trap's characteristics, as well as the laser and control setup are described in [2]. Most notably, this trap possesses a significantly lower heating rate of the axial mode $\dot{\bar{n}} = 1.1(1)$quanta/ s, but also a magnetic field gradient along the trap axis of $0.573(1)$ mG/ μm. While the magnetic field gradient poses a problem for gates between two ions of the same species, because global single-qubit operations will be slightly off-resonant for one or both of the qubits, this is not a problem for different species. Here each ion has its own species-selective microwave or RF single qubit drive, and for fixed ion order they can be adjusted to compensate for the magnetic field offset. The two-qubit gate is not affected because its operation is independent of the qubit frequency. The trap was operated at low field, with $B = 1.95$ G, leading to qubit frequencies of $\omega_{0,^{40}Ca^+} = 2\pi \times 5.46$ MHz and $\omega_{0,^{43}Ca^+} = 2\pi \times 3.220828$ GHz.

© Springer Nature Switzerland AG 2020
V. M. Schäfer, *Fast Gates and Mixed-Species Entanglement with Trapped Ions*,
Springer Theses, https://doi.org/10.1007/978-3-030-40285-3_6

6.1.1 Cooling

The isotope shift, that is the difference of the transition frequency between $^{40}Ca^+$ and centre-of-gravity in $^{43}Ca^+$, is $\Delta f_{istp} = 688(17)$ MHz [3, 4] for the 397 nm transition. The total difference in the transition frequency between $^{40}Ca^+$ and $^{43}Ca^+$ consists of this isotope shift and the hyperfine shifts of the relevant levels. We therefore use two separate 397 lasers for Doppler cooling $^{40}Ca^+$ and $^{43}Ca^+$ respectively, where the 397/43 laser has an additional EOM to address both hyperfine levels. For the 866 nm transition a single laser is used and the isotope splitting is bridged by an EOM. The Raman lasers are a pair of injection-locked frequency-doubled diodes, described in [5], where for this experiment there is no frequency offset between the two lasers. The frequency differences of the beams are therefore only set by AOMs, and Raman sideband cooling and temperature diagnostics are hence only possible for $^{40}Ca^+$. The $^{43}Ca^+$ ion is sympathetically cooled to the ground state. Because of the very similar masses of $^{40}Ca^+$ and $^{43}Ca^+$, motional coupling between the different isotopes is very strong.

6.1.2 State Preparation and Readout

State preparation is performed with a $\hat{\sigma}^+$-polarised beam of the individual 397 lasers. Shelving is done with a single 393 nm laser, with the splitting due to the isotope shift and hyperfine shifts provided by an EOM. For readout we used only one 850 nm laser beam without any sidebands, which is used to create the dark resonance necessary for the $^{40}Ca^+$ EIT state-selective shelving [6]. However the fidelity of this scheme is only ∼90%. The isotope shift for the 850 nm transition is $\Delta f_{istp} = -3462.4(2.6)$ MHz [3]. For $^{43}Ca^+$ the 850 laser is therefore off-resonant and population decayed to $3D_{3/2}$ can not be recovered, reducing the readout fidelity. While shelving of the two species is performed simultaneously, fluorescence collection for the different species is performed isotope selectively in separate time bins. The achieved readout fidelities were LRL = 7.5(3)% and URL = 92.9(3)% for $^{40}Ca^+$, and LRL = 2.0(2)% and URL = 94.7(3)% for $^{43}Ca^+$, corresponding to an average readout error of $\epsilon_{ro} = 5.5\%$. To extract only the error of the gate operation, the readout levels are measured carefully and their effect on the measured populations is inverted, see Sect. 6.1.5. Because the gate errors are much smaller than the readout errors, high statistics in the ROL measurements are necessary and they are measured repeatedly interleaved with gate measurements to reduce the effect of drifts. We used $N = 10.000$ samples for each ROL measurement.

Fig. 6.1 Raman lasers for ^{40}Ca$^+$–^{43}Ca$^+$ gate: A single pair of Raman lasers is used to drive the gate, with a frequency splitting close to the motional frequency of the ions $f_z \approx 2$ MHz. The Raman detuning from the excited $4P_{1/2}$ level is $\Delta = 2\pi \times 1$ THz. Figure published in [1]

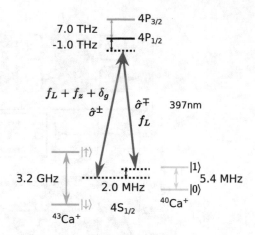

6.1.3 Ion Order

Keeping the correct ion order is important due to the axial magnetic field gradient. Because of the low mass difference between ^{40}Ca$^+$ and ^{43}Ca$^+$, deterministic ordering using the mass dependence of the RF potential—as it is used for ^{43}Ca$^+$ and ^{88}Sr$^+$—is not very effective. Instead we take advantage of the large magnetic field gradient and probe the qubit transition frequency of both qubits with a slow $t \approx 100\,\mu$s carrier π-pulse. If the pulse is off-resonant by an equal frequency offset for both species, the external magnetic field has drifted and a magnetic field servo is applied. If there is a differential change in frequency for the two species, the ion order is incorrect and the crystal is melted using Doppler heating. The ions are then re-crystallised, by switching to the weak trap and Doppler cooling them, followed by returning to the tight trap. This process is repeated until the ions have crystallised in the correct order.

6.1.4 Gate Implementation

The gate is implemented on the ip mode ($f_{\text{ip}} = 1.998$ MHz), with half-integer ion spacing. The Raman laser frequencies relative to the different ion species are shown in Fig. 6.1. Due to the similar masses the Lamb-Dicke factors differ only slightly, with $\eta_{^{43}\text{Ca}^+} = 0.126$ and $\eta_{^{40}\text{Ca}^+} = 0.121$. The gate efficiency is close to one ($\zeta = 0.98$).

Due to the different amplitudes of the light shift force for the different species a single-qubit phase is acquired during the gate operation. To cancel this phase the gate is performed in a two-loop arrangement, see Fig. 6.2. In a gate measurement sequence we first stabilise the Raman laser intensity. The ions are then cooled to the ground state $\bar{n} < 0.1$ first with Doppler cooling followed by sideband cooling for both motional modes. Next, the ions are prepared into $|\downarrow\ 0\rangle$ and the gate is performed within a

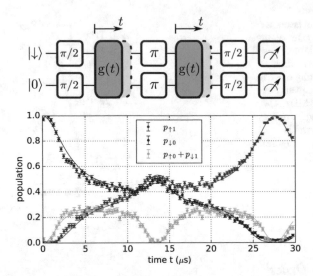

Fig. 6.2 Time scan of a $^{40}Ca^+$–$^{43}Ca^+$ gate with a two-loop arrangement. Top: Pulse sequence used for the measurement. The total length of the gate sequence is $t_{tot} = 2t$ (plus the length of the single qubit operations). We define the pulse duration t such that $t_g = 2\pi/\delta_g$, and therefore a k-loop gate has a total length of $t_{tot} = kt_g$. Bottom: Gate dynamics for increasing length of the gate pulse, after readout level normalisation. The ions' initial state is $|\downarrow\ 0\rangle$, and the populations are gradually transferred towards a Bell state $|\downarrow\ 0\rangle + |\uparrow\ 1\rangle$ at $t = 13.7\,\mu s$. If the gate pulse is applied for a longer time the ions are disentangled again until they reach $|\uparrow\ 1\rangle$ at $t = 27.4\,\mu s$. Gate dynamics time-scans are used for calibration of the gate pulse length and Raman power

spin-echo Ramsey interferometer. Finally both ions are shelved simultaneously, and then their respective state is read out sequentially through fluorescence detection. The single-qubit operations constituting the spin-echo interferometer are driven with microwaves for $^{43}Ca^+$ and with Raman lasers (in a configuration that does not couple to the ion's motion) for $^{40}Ca^+$.

For measuring the gate fidelity an additional $\pi/2$-pulse is added to the sequence, see Fig. 6.3. The contrast of the measured parity fringes is fitted using a maximum likelihood method [7]. We obtain a fidelity of $\mathscr{F} = 99.8(6)\%$ for a gate of total length $t_{tot} = 27.4\,\mu s$.

6.1.5 Bell Test

Violation of the principle of local realism, that is that two space-like separated objects can not influence each other and that each object has a predetermined state independent of measurement, is one of the most mind-bending properties of quantum mechanics. The defiance of local realism in quantum mechanical systems can be proven by violation of the inequality proposed by Bell [8]. In addition, the degree of

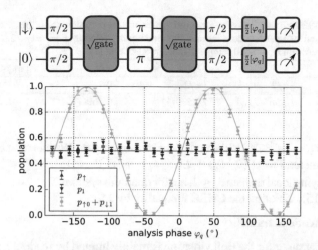

Fig. 6.3 Parity phase scan used to determine the gate fidelity. Top: Pulse sequence used for the parity measurement. The phase φ_q of the final $\pi/2$-pulse at the end of the gate is scanned φ_q : $-180° \mapsto 180°$. Bottom: Measured parity signal (turquoise) and populations (red, ^{43}Ca$^+$; blue ^{40}Ca$^+$). The populations are independent of φ_q. The populations for determining the gate fidelity ($p_{\downarrow 0}$ and $p_{\uparrow 1}$) are measured without the additional $\pi/2$-pulse. Figure published in [1]

violation of Bell's inequality can serve as a measure of entanglement between particles [9]. While the experimental violation of Bell's inequality was shown soon after its proposition [10], the first loophole-free demonstration was only performed recently [11]. Loopholes mean that due to experimental constraints additional assumptions have to be made, which, if incorrect, could explain the violation of Bell's inequality without violating local realism. The most common loopholes are the detection and the locality loophole [12]. The locality loophole exists when a signal could travel between the two measurement sites within the duration of the measurement, as this would allow communication between the two measurement devices that could influence the measurement. The detection loophole exists when the measurement devices have non-unity detection efficiency and therefore in some runs of the experiment a result isn't recorded by either one or both detectors. Hence, while the subset of detected measurements violates the Bell inequality, the set of all runs might not, if the sample of the successful detections is not a 'fair' representation of the whole set. The state of trapped ions can be read out with near unit detection efficiency and errors in the detection are typically low enough that the fair sampling assumption does not have to be made. Indeed the first Bell inequality violation with the detection loophole closed was performed in trapped ions [13].

We measure the violation of the CHSH inequality [14], a modified version of Bell's inequality that is easier to realise experimentally. For this we measure the two-particle correlations $E(\theta_a, \theta_b)$ for different bases, where the basis angles θ_a, θ_b are chosen to obtain maximal CHSH inequality violation for our Bell state, see Table 6.1. The correlations are defined as $E(\theta_a, \theta_b) = p_{\uparrow\uparrow} + p_{\downarrow 0} - p_{\uparrow 0} - p_{\downarrow 1}$ [15], with populations $p_{s1,s2}$ measured after applying the single-qubit rotations $R(\theta_a)$ on

Table 6.1 CHSH Bell test: Sets of angles used to measure the correlations for the CHSH inequality, and the obtained two-particle correlations. For maximum violation $E = \pm 1/\sqrt{2}$. Each value of E was measured with $N = 2000$ samples. Table published in [1]

θ_a (^{40}Ca$^+$)	$\pi/4$	$3\pi/4$	$\pi/4$	$3\pi/4$
θ_b (^{43}Ca$^+$)	$\pi/2$	$\pi/2$	0	0
$E(\theta_a, \theta_b)$	0.565(7)	0.530(7)	0.560(7)	−0.573(8)

^{40}Ca$^+$ and $R(\theta_b)$ on ^{43}Ca$^+$ at the end of the two-qubit gate. We can then calculate the CHSH parameter $S = |E(\theta_a, \theta_b) + E(\theta'_a, \theta_b)| + |E(\theta_a, \theta'_b) - E(\theta'_a, \theta'_b)|$, where for systems obeying local realism $S \leq 2$, but for quantum systems $S \leq 2\sqrt{2}$. We measure $S = 2.228(15)$, violating the CHSH inequality by 15σ.

Effects of Readout Infidelity

The value measured for the Bell violation is mainly limited by readout imperfection, that is an ion in $|\uparrow\rangle$ is falsely registered as $|\downarrow\rangle$ or vice versa. We can model the effect of readout errors with the operator

$$M_i = \begin{pmatrix} \beta_i & \alpha_i \\ 1 - \beta_i & 1 - \alpha_i \end{pmatrix} \tag{6.1}$$

where α_i (β_i) are the measured LRL (URL) listed in Sect. 6.1.2 for ^{40}Ca$^+$ and ^{43}Ca$^+$. The total effect of the readout is then described by applying the matrix $M = M_{^{40}\text{Ca}^+} \otimes M_{^{43}\text{Ca}^+}$ to the vector of real ion populations, $|\psi_{\text{meas}}\rangle = M |\psi_{\text{real}}\rangle$. Here $|\psi_{\text{real}}\rangle = \begin{pmatrix} p_\uparrow \\ p_\downarrow \end{pmatrix}$ and the readout levels are the populations measured for $p_\uparrow$ when preparing the qubit in $|\downarrow\rangle$: $\alpha = p_{\uparrow,\text{meas,min}}$, and $|\uparrow\rangle$: $\beta = p_{\uparrow,\text{meas,max}}$.

We can estimate the maximum value measurable for S with our readout levels by applying M to the density matrix of the ideal Bell state. We obtain $S_{\text{max}} = 2.236(7)$. The value we measure for S is therefore in agreement with the high Bell state fidelity deduced from parity oscillations.

Readout level normalisation for mixed-species experiments, for example used in measurements of the gate fidelity, is performed by applying M^{-1} to the vector of measured populations.

6.2 ^{43}Ca$^+$–^{88}Sr$^+$ Gate Results

Experiments entangling ^{43}Ca$^+$ and ^{88}Sr$^+$ were performed in the new trap and laser setup and with the Ti:Sapph lasers as Raman lasers. The gate mechanism used was identical to the one used for entangling ^{40}Ca$^+$ and ^{43}Ca$^+$. Both gate beams (R_V and $R_\parallel$) are derived from the master laser, while sideband cooling is performed using the Raman beams $R_\parallel$ and R_H derived from the master and phase-locked slave laser respectively. Dark resonance cooling was not yet implemented, leading to very high temperatures on ^{43}Ca$^+$ after Doppler cooling. To reduce the effects of poor Doppler cooling on ^{43}Ca$^+$, first Doppler cooling was performed for $t = 1$ ms on both species simultaneously, followed by Doppler cooling on ^{88}Sr$^+$ only, for $t = 0.5$ ms, during which the ^{43}Ca$^+$ ion is cooled sympathetically. This is followed by sideband cooling, interleaved on both motional modes, for $t = 6$ ms per mode in total. Sideband cooling is performed on ^{43}Ca$^+$, with ^{88}Sr$^+$ cooled sympathetically. At the time of this experiment the blue-blue dark resonance in the 397 nm cooling laser (see Sect. 5.2.1), which also affects sideband cooling, had not yet been identified. This led to very poor cooling performance in spite of the very long cooling times. The final achieved temperature after sbc was $0.2 \lesssim \bar{n}_{\text{oop}} \lesssim 0.5$, measured on ^{43}Ca$^+$.

State preparation is performed simultaneously on ^{43}Ca$^+$ and ^{88}Sr$^+$. For readout we first shelve and readout ^{43}Ca$^+$, followed by ^{88}Sr$^+$.

6.2.1 Choosing the Raman Detuning

The choice of the Raman detuning not only affects the scattering rate on each ion, but also the respective Rabi frequencies and therefore the gate efficiency. Because of the fairly large Raman detunings, the Rabi frequencies on the ion are rather small. Since Raman power is limited, the Raman detuning therefore also affects the minimum possible gate length. Due to the large heating rate of the trap, performing slow gates on the ip mode would make motional heating effects the dominant contribution to the gate error, see Fig. 6.4. All our gates are therefore performed on the oop mode, which has a much smaller heating rate. However, due to the mass asymmetry between ^{43}Ca$^+$ and ^{88}Sr$^+$, the oop heating rate is expected to be large enough to still be the dominant source of error. The Raman detuning was set to $\Delta = -9$ THz from the 397 nm resonance. This corresponds to a gate efficiency $\zeta = 0.81$ for a gate on the oop mode (see Eq. 3.53).

6.2.2 Preliminary Results

We performed a proof-of-principle demonstration of the ^{43}Ca$^+$–^{88}Sr$^+$ gate, with the resulting parity fringes shown in Fig. 6.5. The gate was performed at $t_{\text{gt}} = 104\,\mu$s in a two-loop configuration. The normal mode frequency was $f_{\text{oop}} = 2.851$ MHz. Single-qubit operations were performed using microwaves for ^{43}Ca$^+$, and using RF for ^{88}Sr$^+$. The parity fringes are normalised for readout, which were LRL$_{43\text{Ca}^+} = 1.4\%$, URL$_{43\text{Ca}^+} = 99.4\%$ for ^{43}Ca$^+$ and LRL$_{88\text{Sr}^+} = 15.2\%$, URL$_{88\text{Sr}^+} = 90.3\%$ for ^{88}Sr$^+$ at the time the experiment was performed. For ^{43}Ca$^+$ these readout levels are below their best performance due to imperfect state preparation caused by the blue-blue dark resonance. For ^{88}Sr$^+$ drifts of the locking cavity of the 408 nm shelving laser reduced the LRL fidelity.

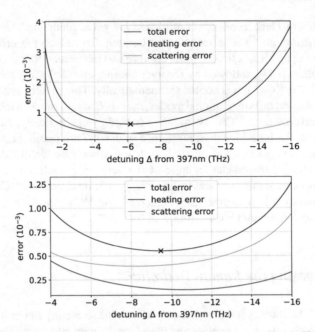

Fig. 6.4 Gate error versus Raman detuning: Simulation of the error of a $^{43}\text{Ca}^+$–$^{88}\text{Sr}^+$ entangling gate as a function of Raman detuning Δ. Top: Gate on the ip mode, assuming a heating rate of $\dot{\bar{n}} = 93$ quanta/s. Bottom: Gate on the oop mode, assuming $\dot{\bar{n}} = 27$ quanta/s. Both simulations assume a Raman spotsize of $r_0 = 25\,\mu\text{m}$ at the ion and $P = 70\,\text{mW}$ in each beam. The black cross indicates the point of minimum total error. The detuning used in the experiment was $\Delta = -9\,\text{THz}$. The values for the heating rates are calculated from the measured heating rate of a single $^{43}\text{Ca}^+$ ion

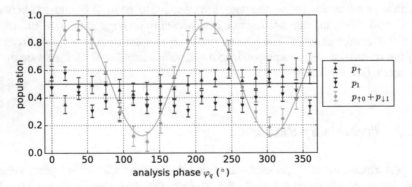

Fig. 6.5 Parity phase scan of $^{43}\text{Ca}^+$–$^{88}\text{Sr}^+$ gate: Measured parity signal (turquoise) and populations (red, $^{43}\text{Ca}^+$; blue $^{88}\text{Sr}^+$). For a perfect Bell state the populations are independent of φ_q. The populations for determining the gate fidelity ($p_{\downarrow 0}$ and $p_{\uparrow 1}$) are measured without the additional $\pi/2$-pulse. The inferred fidelity is $\mathcal{F} = 83(3)\%$

The fidelity of the ^{43}Ca$^+$–^{88}Sr$^+$ gate was $\mathscr{F} = 83(3)\%$. This is believed to have been mainly limited by imperfect cooling. The low fidelity of the preliminary readout scheme and problems with drifting locking cavities, drifting trap compensation and drifts of the laser polarisation axis in optical fibres additionally complicated diagnosis of the source of errors. These problems have all been solved now and high fidelity σ_z gates were performed in this trap with the same Raman lasers on ^{43}Ca$^+$–^{43}Ca$^+$, see Chap. 7. Although we did not yet have time to revisit this experiment, we believe the improvements made should considerably increase the fidelity of this gate.

References

1. Ballance CJ et al (2015) Hybrid quantum logic and a test of Bell's inequality using two different atomic isotopes. Nature 528:384–386. ISSN: 0028-0836
2. Ballance CJ (2014) High-fidelity quantum logic in Ca + PhD thesis, University of Oxford
3. Szwer D (2010) High fidelity readout and protection of a 43Ca+ trapped ion qubit PhD thesis, University of Oxford. papers://d311e016-dabd-41c6-98d5-71ce9eddf36c/Paper/p1856
4. Lucas DM et al (2004) Isotope-selective photoionization for calcium ion trapping. Phys Rev A 69:012711. ISSN: 1050-2947
5. Linke NM, Ballance CJ, Lucas DM (2013) Injection locking of two frequency doubled lasers with 3.2 GHz offset for driving Raman transitions with low photon scattering in 43Ca+. Opt Lett 38:5087–5089
6. McDonnell MJ, Stacey DN, Steane AM (2004) Laser linewidth effects in quantum state discrimination by electromagnetically induced transparency. Phys Rev A 70:053802. ISSN: 10502947
7. Ballance CJ, Harty TP, Linke NM, Sepiol MA, Lucas DM (2016) High-fidelity quantum logic gates using trapped-ion hyperfine qubits. Phys Rev Lett 117:060504. ISSN: 10797114
8. Bell JS (1964) On the Einstein Podolsky Rosen paradox. Physics 1:195–200
9. Lanyon BP et al (2014) Experimental violation of multipartite Bell inequalities with trapped ions. Phys Rev Lett 112:100403. ISSN: 0031-9007
10. Freedman SJ, Clauser JF (1972) Experimental test of local hidden-variable theories. Phys Rev Lett 28:938–941
11. Hensen B et al (2015) Loophole-free Bell inequality violation using electron spins separated by 1.3 kilometres. Nature 526:682–686. ISSN: 0028-0836
12. Brunner N, Cavalcanti D, Pironio S, Scarani V, Wehner S (2014) Bell nonlocality. Rev Mod Phys 86:419–478. ISSN: 15390756
13. Rowe MA et al (2001) Experimental violation of a Bell's inequality with efficient detection. Nature 409:791–794
14. Clauser JF, Horne MA, Shimony A, Holt RA (1969) Proposed experiment to test local hidden-variable theories. Phys Rev Lett 23:880–884. ISSN: 03759601
15. Matsukevich DN, Maunz P, Moehring DL, Olmschenk S, Monroe C (2008) Bell inequality violation with two remote atomic qubits. Phys Rev Lett 100:150404. ISSN: 0031-9007

Chapter 7
Fast Gates

The speed of the trapped ion entangling gates described in Sect. 3.4 can be increased by using a larger gate detuning δ_g, while at the same time increasing the Raman intensity to maintain the acquired geometric phase $\Phi = \pi/2$. This will lead to several error sources becoming more relevant:

Off-resonant excitation: For shorter gate pulses errors due to off-resonant excitation increase. The sharp edges of the pulse have a broad frequency spectrum. These different frequency components can drive the second motional mode, higher orders and counter-rotating modes of the main motional mode and will cause an AC Stark shift coupling to the carrier depending on the un-stabilised relative phase of the Raman beams ϕ_0. Shaping the edge of the pulse on the timescale of a few motional periods of the ion's motion reduces the pulse bandwidth and strongly suppresses errors due to off-resonant excitation [1]. However this method starts to fail once the total gate length is on the same order of magnitude as the ideal pulse shaping length.

Raman Scattering: Faster gates require larger Rabi frequencies. Because laser-power is limited, this means smaller Raman detunings might be necessary to obtain sufficient Rabi frequency. This will lead to larger scattering errors. Although a larger Raman power with constant Raman detuning won't lead to an increase of scattering errors (thanks to the decreased gate time), the reduced gate efficiency for faster gates will still lead to an increase of scattering errors.

Rotating wave approximation failure: Failure of the rotating wave approximation causes the geometric phase to become sensitive to the Raman phase ϕ_0. Other motional modes are not only excited by frequency components due to the sharp edges of the pulse, but also because their detuning to the Raman beat note is now of similar magnitude to the detuning from the desired mode.

Therefore it is necessary to implement special gate sequences that are designed to avoid the above mentioned sources of error, as discussed in Sect. 3.6.

We implement 3 different kinds of solutions:

(1) In the intermediate time regime (ii) (see Sect. 3.6.3) down to about 4 motional periods solutions can be found using single rectangular unshaped pulses. Here

© Springer Nature Switzerland AG 2020

V. M. Schäfer, *Fast Gates and Mixed-Species Entanglement with Trapped Ions*,
Springer Theses, https://doi.org/10.1007/978-3-030-40285-3_7

the pulse power and detuning are optimized for fixed times to close loops of both motional modes. For shorter gate times errors increase and no more useful solutions can be found.

(2) Binary solutions consist of a sequence of short pulses interleaved with delays of free evolution. The amplitude level of all pulses is the same, making the sequences easy to implement and easier to optimise due to the reduced number of degrees of freedom. However, while the integrated Rabi frequency is lower due to the delays with no driving force, the required peak Rabi frequency is considerably higher compared to gates of type (3) with the same total gate length.

(3) Stepped pulse sequences consist of a single symmetric pulse with up to four different amplitude levels. The majority of our implemented fast gate sequences is of this type.

7.1 Experimental Details

7.1.1 Experiment Setup

The experimental setup is described in Fig. 7.1. One measurement shot begins with intensity stabilisation of the Raman lasers. It is followed by ground state cooling of the ions, consisting of dark-resonance Doppler cooling, followed by continuous sideband cooling and pulsed sideband cooling—each for both motional modes. The final temperatures are typically $\bar{n}_{ip} \leq 0.05$ for the in-phase com mode and $\bar{n}_{oop} \leq 0.02$ for the out-of-phase stretch mode. Next the ions are prepared in $|\downarrow\downarrow\rangle$ after which the main experimental sequence follows. The two-qubit gate is placed inside a Ramsey interferometer. The carrier pulses of the Ramsey interferometer are driven by microwaves. After readout of the final state we test whether both ions are still crystallized. If not, an automatic re-crystallization occurs and the last data point is discarded.

In early measurements we noted a dependence of the gate fidelity on the 866 detuning. This is because the dark-resonance cooling sensitively depends on the 866 frequency (see Fig. 5.5) and a mis-set of this frequency can leave population in very high-n states. To avoid this, cooling times in the later fast gate experiments were chosen very generously. Subsequently no decrease in fidelity was noted, even for intentional mis-set of the 866 frequency by a multiple of the typical drift.

7.1.2 Measuring Gate Fidelities

The fidelity of the gate-operations is determined via the Bell state fidelity, as in [3]. The phase-gate is placed inside a Ramsey interferometer, see Fig. 7.2, to produce the ideal Bell state $|\psi_{Bell}\rangle = \frac{1}{\sqrt{2}} (|\uparrow\uparrow\rangle + i |\downarrow\downarrow\rangle)$. The fidelity of the created state is then measured by partial tomography:

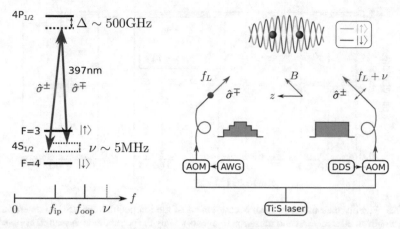

Fig. 7.1 Left: Level structure: The gate is performed on the 'stretch'-qubit in $^{43}Ca^+$, comprising the two qubit levels $|\downarrow\rangle = |F = 4, m_F = 4\rangle$ and $|\uparrow\rangle = |F = 3, m_F = 3\rangle$. Depending on the required Rabi frequency the Raman detuning is varied between $\Delta = 200\,GHz - 1\,THz$. The gate mechanism is based on the geometric phase gate and the difference frequency of the two Raman lasers is therefore independent of the qubit frequency, and only depends on the motional mode frequencies of the ions $f_{ip} = 1.92\,MHz$ and $f_{oop} = 3.33\,MHz$. In contrast to slow gates the gate detuning is large enough that it couples to both modes. For the implemented gate sequences the beatnote frequencies were between $\nu = 2.0\,MHz - 7.5\,MHz$. Right: Beam geometry: Both Raman lasers are derived from a single cw Ti:Sapphire laser. Both beams are modulated with fast AOMs of risetime $t_r \approx 24\,ns$ to form short pulses. An AWG provides the RF for the more complex pulse shape of one beam, the other RF pulse is a simple square pulse produced with a DDS board. Fibres are used to produce a clean spatial and temporal mode profile of the beams and to reduce beam-pointing fluctuations at the ions. The two Raman beams interfere to form a travelling standing wave along the trap axis. The light field gives rise to a spin-dependent force on the ions' motion, thus driving the gate dynamics. Figure published in [2]

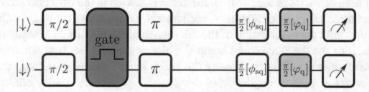

Fig. 7.2 Gate sequence: The ions are prepared in $|\downarrow\downarrow\rangle$. The gate pulse is placed in one arm of a spin-echo sequence. The second $\pi/2$-pulse of the Ramsey interferometer is offset by a phase ϕ_{sq} to compensate for the single-qubit phase accumulated during the gate. For measuring the x- and y-quadratures of the parity a second $\pi/2$ pulse is added at 4 different phases $\varphi_q = 0$, $\pi/4$, $\pi/2$, $3\pi/4$. A spin-echo sequence is used for suppression of magnetic field noise. For fast sequences a spin-echo sequence is no longer necessary, but was used for better comparability to our other gate measurements

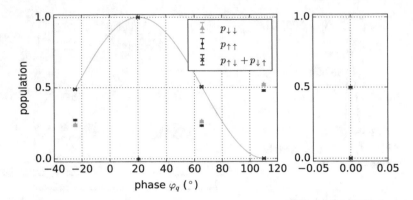

Fig. 7.3 Fidelity measurement: Left: Measurement of the ion populations used to determine the parity contrast, after readout normalisation (magenta). Only the populations measured at $\varphi_q = 20°$ and $\varphi_q = 110°$ are used to estimate the fidelity. The populations measured in the orthogonal basis and the grey line are only added for clarity. Right: Measurement of the populations without the analysis pulse

$$\mathscr{F} = \langle \psi_{\text{Bell}} | \rho | \psi_{\text{Bell}} \rangle = \frac{1}{2} \text{ (population + parity)} \tag{7.1}$$

The population $= p_{\downarrow\downarrow} + p_{\uparrow\uparrow}$ can be measured directly at the end of the Ramsey sequence. The parity is the contrast c of the parity fringes $\mathscr{P}(\varphi_q) = p_{\downarrow\downarrow} + p_{\uparrow\uparrow} - p_{\uparrow\downarrow} - p_{\downarrow\uparrow} = c \cdot \sin\left(2\varphi_q + \varphi_0\right)^1$. It is measured by adding an additional $\pi/2$-pulse after the Ramsey sequence and scanning its phase φ_q.

For single rectangular pulses we measure the parity for 4 different phases $\varphi_q = (0, \pi/4, \pi/2, 3\pi/4)$ and determine the contrast with a maximum likelihood fit [1, 4]. The phase offset φ_0 of the parity oscillations is floated in the fit. This means noise in the measured parity quadratures can lead to a mis-estimation of the phase-offset and therefore a slight over-estimation of the fidelity. To get a more reliable measurement of the fidelity for the binary and stepped pulse sequences we first determine the phase-offset φ_0 and then only measure the parity in the basis of maximum contrast, at $\varphi_{q,1} = \pi/4 + \varphi_0$ and $\varphi_{q,2} = 3\pi/4 + \varphi_0$, see Fig. 7.3. The parity contrast can then be calculated directly as $c = \frac{1}{2}\left(\mathscr{P}(\varphi_{q,1}) - \mathscr{P}(\varphi_{q,2})\right) = \left[p_{\downarrow\uparrow} + p_{\uparrow\downarrow}\right]_{\varphi_{q,1}} - \left[p_{\downarrow\uparrow} + p_{\uparrow\uparrow}\right]_{\varphi_{q,2}}$, without any fitting routines necessary.

7.1.2.1 Readout Normalisation

A complete quantum algorithm consists of many gate operations, but only a single instance of state-preparation and readout for each qubit. Therefore it is desirable to distinguish errors stemming from the gate operation and errors occurring

---

[1]Note the difference between ϕ_0, the initial relative phase of the Raman lasers and φ_0, used for various other phase offsets.

during SPAM. We measure the readout levels interleaved with the gate fidelity measurements. The gate fidelity data is then corrected for readout errors by applying $|\psi_{\text{corr}}\rangle = M^{-1} |\psi_{\text{meas}}\rangle$, where the readout level matrix is

$$
M = \begin{pmatrix} \beta^2 & \alpha\beta & \alpha^2 \\ 2\beta(1-\beta) & \beta(1-\alpha) + \alpha(1-\beta) & 2\alpha(1-\alpha) \\ (1-\beta)^2 & (1-\beta)(1-\alpha) & (1-\alpha)^2 \end{pmatrix} \tag{7.2}
$$

In contrast to the mixed species readout in Sect. 6.1.5 we cannot distinguish between $p_{\uparrow\downarrow}$ and $p_{\downarrow\uparrow}$. As a result the measured probability vector is $|\psi_{\text{meas}}\rangle = (p_{\uparrow\uparrow,\text{meas}}; \; p_{\uparrow\downarrow,\text{meas}} + p_{\downarrow\uparrow,\text{meas}}; \; p_{\downarrow\downarrow,\text{meas}})$ and we accordingly sum together the mixed terms in M [1]. All fidelities quoted are after readout normalisation.

7.2 Gate Sequences

7.2.1 Single Rectangular Pulses

The most simple gate sequences consist of a single rectangular pulse, where the only free parameters are the gate detuning $\delta_g = \nu - f_{\text{ip}}$, the gate length t_g and the Raman power P. Solutions for single rectangular pulses are found by simulating a gate with fixed t_g and then minimising the coherent gate-error with a Nelder-Mead algorithm, while floating the parameters δ_g and P. The simulated parameters are then further optimized empirically: they are dialled into the experiment and the measured gate fidelity is fed back to a Nelder-Mead algorithm to return new optimised parameters.

Apart from the spin-dependent force the laser field also leads to a time-varying light shift, that causes a single-qubit phase ϕ_{LS}. Because the matrix elements describing the Raman coupling to the two qubit levels $|\uparrow\rangle$ and $|\downarrow\rangle$ are not identical, the force acting on $|\uparrow\rangle$ is different to the force acting on $|\downarrow\rangle$. This leads to a geometric single qubit phase ϕ_{gs}. The total single-qubit phase $\phi_{\text{sq}} = \phi_{\text{gs}} + \phi_{\text{LS}}$ is compensated for by adding a phase offset $-\phi_{\text{sq}}$ to the second $\pi/2$-pulse of the Ramsey interferometer. This offset is also optimised empirically by the Nelder-Mead algorithm.

Gate sequences consisting of a single rectangular pulse between $t_g = 1.21\,\mu\text{s}$ and $t_g = 5.96\,\mu\text{s}$ were implemented. Example gate parameters for the fastest and the slowest gate are shown in Table 7.1. The results are shown in Fig. 7.4. Each fidelity measurement consists of 2000 shots of the experiment.

7.2.2 Binary Pulse Sequences

One possibility to introduce new degrees of freedom in the gate pulse is to modulate it as a sequence of short pulses interleaved with delays of free evolution. We have implemented one such sequence with gate parameters outlined in Fig. 7.5.

Table 7.1 Gate parameters for the fastest and slowest gate sequences implemented with a single rectangular pulse. The parameters are the results of empirical optimisation of the theoretically-predicted parameters. The Raman detuning is $\Delta = -1\,\text{THz}$. The gates are driven on the axial ip mode with $f_z = 1.86\,\text{MHz}$, and the power P is equal in both beams

Gate sequence	2 mp	11 mp
Gate time t_g	1.21 μs	5.97 μs
Gate detuning δ_g	756.0 kHz	172.15 kHz
Beam power P	76.44 mW	21.79 mW
Phase ϕ_{sq}	−8.5°	−4.6°

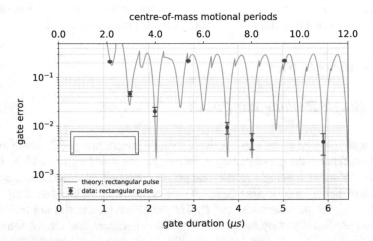

Fig. 7.4 Intermediate speed gate results: Our experimental results fit well to the theoretical predictions. Fidelities ≥98% are measured down to 4 mp, with a maximum of $\mathcal{F} = 99.5(2)\%$ at 8 mp. For smaller gate times we leave the adiabatic regime and errors increase drastically. The risetime of the pulses is limited by the AOM to 24 ns, far below the motional period of the ions; thus the pulses are effectively rectangular. Figure published in [2]

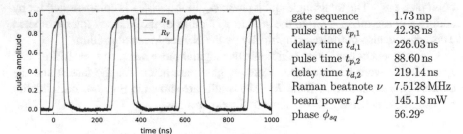

gate sequence	1.73 mp
pulse time $t_{p,1}$	42.38 ns
delay time $t_{d,1}$	226.03 ns
pulse time $t_{p,2}$	88.60 ns
delay time $t_{d,2}$	219.14 ns
Raman beatnote ν	7.5128 MHz
beam power P	145.18 mW
phase ϕ_{sq}	56.29°

Fig. 7.5 Binary gate sequence: Left: Oscilloscope trace of the final binary gate sequence. The pulse sequence is symmetric about its centre. The bulges and notches directly after switching the pulse are caused by pulse-chirping in the AOM, see Sect. 7.3.2, and were mitigated in later experiments. The R_V-pulse is padded by the $R_\parallel$-pulse to reduce effects of timing-jitter between the two pulses as well as uneven amplitude after switching. Right: Final empirically-optimised parameters of the binary gate sequence. The power P is equal in both beams

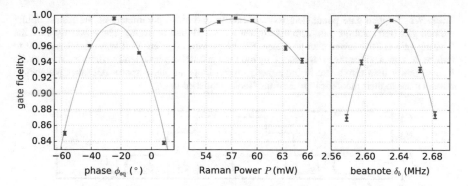

Fig. 7.6 Parameter optimisation: Linear scans of individual parameters are used to optimize stepped gate sequences. The turquoise line is a quadratic fit to the data used to determine the optimal value of the scanned parameter. The above data was measured while optimizing the 3.05 mp gate. Sensitivity to parameters increases for faster gate sequences

The gate sequence is empirically optimized with a Nelder-Mead algorithm, starting from the theoretically predicted values. To improve performance, gate parameters are floated in smaller interdependent groups, such as the power and single qubit phase (P, ϕ_{sq}), the pulse and delay lengths $(t_{p,i}, t_{d,i})$ and pulse lengths combined with the Raman beatnote ν. In a final optimisation all parameters are floated simultaneously. In initial experiments the Raman beam power P is reduced to decrease effects of imperfect loop closure and facilitate optimisation of other parameters by broadening the areas of higher fidelities. In later optimisation series a different amplitude level of the two outer and inner pulses is allowed. However this brings only minimal improvement (Fig. 7.6).

Although binary pulse sequences have fewer degrees of freedom and can be implemented with a simple DDS due to their binary pulse shape we have only realised one such sequence. This is because they require larger peak Raman powers for a given gate time. They are also more sensitive to undesired effects during amplitude switching in the AOM, see Sect. 7.3.2.

7.2.3 Stepped Pulse Sequences

Our fastest gates consist of a single pulse, which has 7 symmetrically arranged segments of varying amplitude levels, see Fig. 7.7. Only the R_V beam is shaped with stepped amplitudes, the $R_\parallel$ beam is left as a rectangular pulse shape, bracketing the R_V at the beginning and end of the pulse. The reason for this is again to be less sensitive to time-jitter between the two pulses as well as to phase-chirps and pulse-shape deviations during and after large RF amplitude steps (see Sect. 7.3.2). The ratio of the peak power of the two beams is chosen to minimize scattering errors. The RF for the $R_\parallel$ AOM is provided by a DDS board, while the RF of the R_V AOM

Table 7.2 Stepped gate parameters Experimentally optimized parameters of the fastest implemented gate (left) and the gate with the highest measured fidelity (right). The power used for the
0.89 mp gate is 20% lower than predicted. For gates with large errors it was often beneficial to
reduce the power, since the errors caused by too little geometric phase tend to increase slower with
Raman power than errors due to imperfectly closed loops in phase space. All other parameters
remain close to their theoretical predictions. Table published in [2]

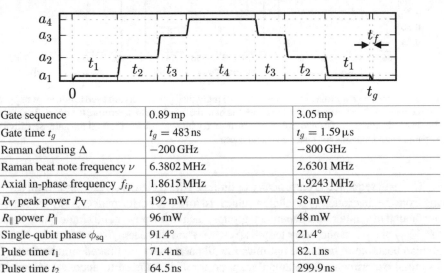

Gate sequence	0.89 mp	3.05 mp
Gate time t_g	$t_g = 483$ ns	$t_g = 1.59\,\mu$s
Raman detuning Δ	-200 GHz	-800 GHz
Raman beat note frequency ν	6.3802 MHz	2.6301 MHz
Axial in-phase frequency f_{ip}	1.8615 MHz	1.9243 MHz
R_V peak power P_V	192 mW	58 mW
$R_\parallel$ power $P_\parallel$	96 mW	48 mW
Single-qubit phase ϕ_{sq}	91.4°	21.4°
Pulse time t_1	71.4 ns	82.1 ns
Pulse time t_2	64.5 ns	299.9 ns
Pulse time t_3	46.7 ns	–
Pulse time t_4	112.3 ns	819.5 ns
Pulse fall-time t_f	5.0 ns	5.0 ns
Pulse amplitude a_1	0.284	0.445
Pulse amplitude a_2	0.617	0.838
Pulse amplitude a_3	0.862	–
Pulse amplitude a_4	1	1

is controlled with the AWG. The phase between the two RF sources is not locked.
The gate parameters for two different sequences are shown in Table 7.2.

Stepped sequences were measured in two different groups. Sequences with gate
times faster than $t_g = 1\,\mu$s were optimised in an analogous way to the binary
sequence, with the added complexity of having 3 different relative amplitude levels a_i.

Sequences with gate times slower than $t_g = 1\,\mu$s were taken after various error
sources had been identified and removed. This means that theoretically-predicted
pulse lengths and amplitudes led to optimal fidelities and could not be improved
further. For these gates the measured fidelities were above $\mathscr{F} = 99\%$. Therefore it is
necessary to collect very large datasets to reduce statistical errors far enough that the
Nelder-Mead optimisation converges to the true optimum and is not dominated by

statistical noise. To improve fidelities above 99% it is faster and yields better results to instead scan individual parameters. For small gate errors all parameters contribute quadratically to the infidelity. The gate parameters P, ν and ϕ_{sq} were all optimised by setting them to the centre of a quadratic fit of the measured gate error. Example scans are shown in Fig. 7.6.

The simulated gate efficiency is $\tilde{\zeta} = 1.234$ for an 11 mp rectangular gate, $\tilde{\zeta} = 7$ for the 3.05 mp gate and $\tilde{\zeta} = 0.14$ for the 0.89 mp gate.

7.3 Error Analysis

Multiple sources of error for the fast gates were modelled and experimentally characterised. Errors due to out-of-Lamb-Dicke effects turned out to be dominant for the fastest gates. The fidelities of the best gates around 3 mp approach the limit of what can be characterised with our measurement method. Since the SPAM errors are of similar size to the gate error more careful characterisation could be done using two-qubit randomised benchmarking. Single-qubit errors of the pulses creating the Ramsey interferometer were not characterised independently and are included in the quoted gate errors.

7.3.1 Pulse Control

The implementation of fast gates requires precise control of the pulse parameters. Due to the non-linear response of the AOM for different RF drive amplitudes the amplitude levels cannot simply be programmed into the AWG but have to be adjusted to give the correct output amplitude. The pulse shape is measured on a photodiode and the relative drive amplitudes are adjusted iteratively to match the measured amplitudes to their theoretically predicted optimum level. To determine the stability of the relative amplitude levels, 50 nominally identical pulse sequences were measured on a photodiode and then fitted. The amplitude levels showed a standard deviation of 0.2%. The AWG used to produce the RF that defines the shape of the R_V pulse has a clock frequency of 1.25 GHz, limiting the timing resolution of the pulses to 0.8 ns. A Monte-Carlo simulation shows that a 0.8 ns standard deviation on the individual pulse lengths in combination with the measured 0.2% amplitude uncertainty leads to an $\epsilon_{gate} \approx 1 \times 10^{-3}$ error in the 3.05 mp sequence and to an $\epsilon_{gate} \approx 3 \times 10^{-2}$ error in the 0.89 mp sequence. To increase the control over the timing precision of the pulses, we programmed a 5 ns sinusoidal risetime for the waveform into the AWG. This is below the 24 ns risetime of the AOM and therefore only negligibly increases the risetime of the Raman pulses at the ion. However it spreads the rising edge over several time points and hence increases control over the timing precision of the rising and falling edges. Fits of 50 measured pulse sequences show that we achieve a 0.2 ns accuracy (standard deviation) of pulse length control. This corresponds to errors of $\epsilon_{\text{gate}} \approx 2(2) \times 10^{-4}$ for the 3.05 mp gate and $\epsilon_{\text{gate}} \approx 1(1) \times 10^{-3}$ for the fastest 0.89 mp gate. The large uncertainties of the errors arise because different combinations of errors on the individual t_i and a_i lead to different magnitudes of total gate error.

The AOM used for the $R_{\parallel}$ beam produces larger beam shape anomalies when switching the amplitude compared to the R_V-AOM. To reduce the effects of these irregularities the length of the $R_{\parallel}$ pulse is increased so that the pulse is switched on (off) before (after) the R_V beam. The length of these paddings is chosen to move all anomalies outside of the R_V window, such that no force will be applied on the ions during that time. The padding also eliminates the effects of relative timing jitter between the two pulses. The increase in scattering error due to the longer exposure on the ions is small compared to the total gate error. To test the choice of the padding length the fidelity was measured with slightly lengthened and shortened padding. The measured fidelity was not affected.

7.3.2 Phase Chirp

AOMs can cause phase chirps in the modulated beams during switching of the RF amplitude. This has been studied in the context of atomic clocks [5]. We characterise the phase-chirps in our system with a homodyne measurement similar to [5]. The measurement setup is illustrated in Fig. 7.8. The phase of the two Raman beams is stabilised in a slow servo-loop to $\delta\phi = \pi/2$, so that the beatnote voltage depends linearly on the phase difference of the two beams. The relative phase between the two lasers is then swapped by π to switch the sign of the effects of phase fluctuations

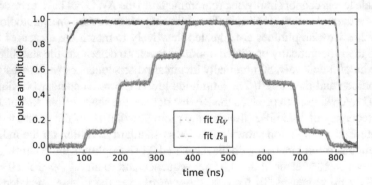

Fig. 7.7 Pulse control: Traces of 50 consecutively measured pulses plotted on top of each other. All sequences are triggered on the rising edge of the $R_{\parallel}$ pulse. The timing jitter between the two pulses visible in the broadened rising edges of the R_V pulse does not affect the gate due to the time padding by the $R_{\parallel}$ pulse. Fits to the data show a 0.2 ns accuracy of the pulse lengths and a 0.2% accuracy of the relative amplitude levels

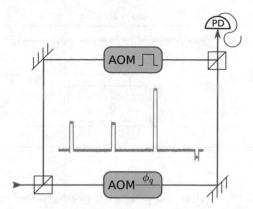

Fig. 7.8 Phase chirp measurement: The laser beam is split into two paths and modulated by two different AOMs. The lower AOM is running constantly but with switched phase-offsets ϕ_q. The upper AOM runs at a constant phase and produces four 500 ns pulses separated by delays. The beams are then superimposed and the beatnote signal is measured on a photodiode. The inset shows the beatnote signal for four pulses. In the delays between the pulses the phase of the lower AOM is switched to $\phi_q = \pi/2, 3\pi/2, 0, \pi$. The baseline corresponds to the photodiode voltage when only the lower beam is switched on and the contrast is limited by imperfect beam-overlap of the two beams on the photodiode

on the beatnote voltage. For reference the points of total constructive and destructive interference are also measured.

The phase chirp can be deduced by subtracting the beatnote pulses measured at $\delta\phi = \pi/2$ and $\delta\phi = 3\pi/2$. The timing of the pulses is determined from simultaneously triggered diagnostic TTL pulses. The results of the measurement are displayed in Fig. 7.9.

The original operating frequency of the AOM $f_{RF} = 217\,\text{MHz}$ was chosen to optimize diffraction efficiencies in all AOMs in the Raman beampath necessary for two-qubit gates, single-qubit gates and sideband cooling. However simulations of the gate error with the phase-chirp produced at this frequency setting show errors of $\epsilon_{\text{gate}} \approx 8 \times 10^{-3}$ for the 3.05 mp gate and $\epsilon_{\text{gate}} \approx 1.3 \times 10^{-1}$ for the 0.89 mp gate sequence. Measurements of the phase-chirp for different RF drive frequencies showed that the chirp was smallest when the AOM was driven at its centre frequency $f_{RF} = 200\,\text{MHz}$. Simulations of errors due to the remaining phase-chirp predict $\epsilon_{\text{gate}} \approx 4(5) \times 10^{-4}$ for the 3.05 mp gate and $\epsilon_{\text{gate}} \approx 6(6) \times 10^{-3}$ for the 0.89 mp sequence. The large uncertainty of the error is due to the imprecise measurement of the magnitude of the remaining phase chirp.

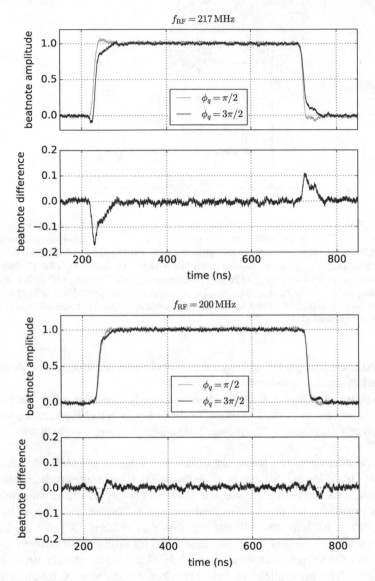

Fig. 7.9 Phase chirp measurements: Top: Phase chirp measured at the original operating frequency of the AOM $f_{RF} = 217\,\mathrm{MHz}$. During the switching of the pulse the quadrature amplitude P_q reaches up to 15% of the in-phase amplitude P_i. Bottom: Operating the AOM at its nominal centre frequency $f_{RF} = 200\,\mathrm{MHz}$ greatly reduces the phase-chirp. Part of the remaining difference signal is due to jitter in the relative pulse timings. This was confirmed by repeating the experiment with the continuously running beam switched off. This way the amplitude of the two pulses is measured and subtracted with the same method, without any beating processes present

Fig. 7.10 Radial mode
errors: Simulated errors due
to radial mode excitation
before (blue) and after (red)
re-alignment of the Raman
lasers. Dips in the errors
arise for sequences where
radial mode phase-space
trajectories are almost closed
at the end of a gate sequence.
The simulations exclude
ooLD effects

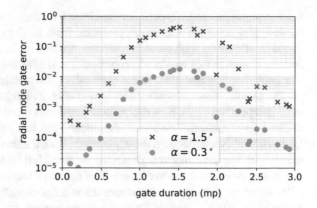

7.3.3   Coupling to Radial Modes

The Raman beatnote frequencies necessary for sequences of around 1.5 mp gate duration lie close to resonance with the radial mode frequencies. Slight misalignment of the Raman lasers with respect to the longitudinal axis of the trap will therefore excite radial motion. The gate sequences are not designed to close trajectories of the radial modes. This leads in general to residual entanglement of the spin with motion in the radial direction at the end of a gate sequence. Because the radial modes are not cooled beyond the Doppler limit this can lead to large errors, see Fig. 7.10.

The differential wave-vector $\Delta\mathbf{k}$ of the two Raman lasers was initially misaligned by $\alpha = 1.5°$ from the trap axis $\hat{\mathbf{z}}$. This leads to gate errors above 40% for the most sensitive gates. After realignment of the Raman lasers no excitation of the radial mode sidebands was visible any more. Changes in the axial mode Lamb-Dicke parameter suggest the mis-alignment was reduced below $\alpha = 0.3°$. This limits errors due to radial mode excitation to $\epsilon_{\text{gate}} \lesssim 2 \times 10^{-2}$, far below errors due to out-of-Lamb-Dicke effects experienced for these gate durations.

After re-alignment of the Raman lasers the axial frequency of the trap had to be changed from $f_{\text{ip}} = 1.86\,\text{MHz}$ to $f_{\text{ip}} = 1.92\,\text{MHz}$ to maintain the ion spacing of $12.5\lambda_{\hat{z}}$. Here $\lambda_{\hat{z}}$ is the periodicity of the travelling standing wave providing the gate force, and depends on the projection of $\Delta\mathbf{k}$ on the trap axis $\hat{\mathbf{z}}$.

7.3.4   Out-of-Lamb-Dicke Effects

The initial relative phase ϕ_0 between the two Raman lasers providing the gate force is not controlled in our experiment. Although both beams originate from the same laser, they do not propagate along the same paths. This is necessary because for being able to couple to the motion of the ions, the beams' differential wave-vector $\Delta\mathbf{k}$ has to be non-vanishing. Small vibrations of the mirror mounts as well as air draughts cause the relative phase to wander on a timescale of ~ 0.1 s. Phase fluctuations could be reduced by designing compact beam paths protected from air draughts, and the relative phase could be stabilised with an active servo-loop feeding back to the RF source of the AOMs. However this would add complexity and slow down the experimental cycle. For implementing fast gates this is especially undesirable.

The gate sequences developed by Steane *et al.* and implemented here are designed to be independent of ϕ_0. However all considerations made in [6] are within the Lamb-Dicke approximation. This means the extent of an ion's wavefunction is assumed to be considerably smaller than the wavelength λ_z of the travelling standing wave interacting with the ions' motion. Therefore the force felt by an ion due to the light-field is constant over all of the spatial extent of the ion's wavefunction. Unfortunately for gates of length $\lesssim 2$ mp the displacement **x** is large enough that this assumption starts to crumble. Instead, depending on its displacement, the force the ion experiences changes, because it interacts with a different part of the standing wave light-field. Because the displacement and force both strongly depend on the initial phase ϕ_0, the amplitude and phase of the force modulation due to out-of-Lamb-Dicke effects strongly depends on ϕ_0. The gate is no longer insensitive to ϕ_0. This leads to large errors.

Figure 7.11 shows the ion's trajectory for a gate sequence where the Lamb-Dicke approximation is still valid. Independently of the phase ϕ_0 the ions' motion returns to its origin at the end of the sequence. This is no longer valid outside of the Lamb-Dicke regime, as shown in Fig. 7.12. Even though a sequence can be designed that returns the ions to their origin for a specific phase, this won't be the case for an arbitrary phase. Averaged over the many shots taken for determining the gate fidelity this results in a large error. Although we can identify sequences shorter than 2 mp that have coherent errors $\epsilon_{\text{gate}} < 1 \times 10^{-2}$, these gate sequences are far less efficient compared to our typically implemented sequences and require considerably more Raman laser power. No sequences with coherent errors $\epsilon_{\text{gate}} < 1 \times 10^{-3}$ have been identified below 2.5 mp, see Fig. 7.13.

The numerical simulations outside of the Lamb-Dicke regime were performed by C.J. Ballance.

7.3.5   Micro-Motion

The original design of the trap [7] includes a large micro-motion gradient along the axial direction so that ions can be addressed via their micro-motion. For two ions with ion spacing $\delta z \approx 3.5\,\mu$m the Rabi-frequency of the first micro-motion sideband is about $1/4$ times the carrier Rabi frequency. The frequency of the micro-motion $f_{\text{mm}} = 28$ MHz is detuned far enough from the gate frequency that it does not significantly affect the gate dynamics. However the Raman powers have to be increased slightly. Simulations outside of the Lamb-Dicke regime including micro-motion show that errors caused by micro-motion are about one order of magnitude smaller than out-of-Lamb-Dicke errors.

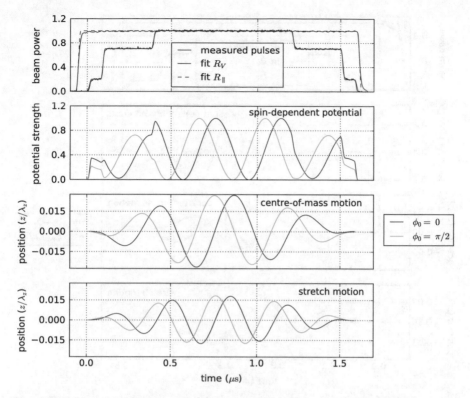

Fig. 7.11 3.05 mp gate sequence: The plot on top shows the pulse amplitudes of the R_V and $R_{\parallel}$ beams together with a fit to the data. The second plot shows the simulated beatnote of the two beams, for two different initial phases ϕ_0. This beatnote corresponds to the spin-dependent potential that gives rise to the force that drives the gate. The lower two plots show the displacements of the ip and the oop mode. Although the forces for different initial phases are decidedly different, and therefore also the trajectories in phase space, the displacements of both modes nevertheless finish at zero displacement at the end of the sequence independent of ϕ_0. Figure published in [2]

7.4 Fast Gate Results

The results for stepped and binary gate sequences are shown in Fig. 7.14. The best fidelity achieved is $\mathscr{F} = 99.78(3)\%$ for the 3.05 mp sequence. This value is an average of 5 different data-points of 10.000 gates each, taken on two different days. This corresponds roughly to a total of 30 min data acquisition time. Directly after calibration of the gate sequence parameters the highest measured gate fidelity was 99.85(3)%. The fidelity decreased to 99.72(3)% after one hour. All fidelities are corrected for SPAM errors. These are measured in interleaved experiments with N = 50.000 shots per readout level and were $\epsilon_{|\downarrow\rangle} = 9(2) \times 10^{-4}$ and $\epsilon_{|\uparrow\rangle} = 1.5(1) \times 10^{-3}$ when this data was taken. The raw fidelity before SPAM normalisation is

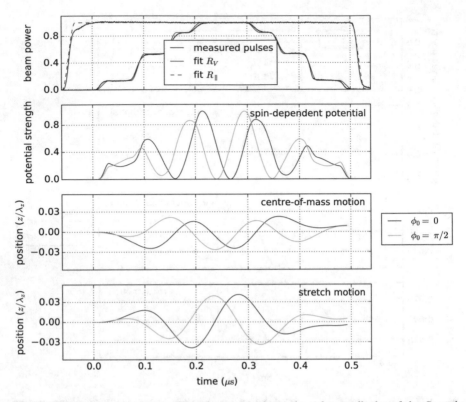

Fig. 7.12 0.89 mp gate sequence: The plot on top shows the pulse amplitudes of the R_V and $R_{\parallel}$ beams together with a fit to the data. In comparison to the 3.05 mp sequence phase chirps lead to small asymmetries in the pulse, causing deviations between fit and data. The second plot shows the simulated beatnote of the two beams, for two different initial phases ϕ_0. The lower two plots show the displacements of the ip and the oop mode. Out-of-Lamb-Dicke effects compromise the phase-independence of the gate sequence leading to differing and non-zero displacements at the end of the sequence

$\mathcal{F} = 99.36(3)\%$. The fastest gate sequence—0.89 mp—is only $t_g = 483$ ns long and has a fidelity of $\mathcal{F} = 60(5)\%$.

Error budgets for the highest-fidelity and for the fastest gate are shown in Table 7.3. While the 0.89 mp gate error is clearly dominated by out-of-Lamb-Dicke errors, this is no longer the case for the 3.05 mp sequence.

Data-points with $t_g < 1\,\mu$s were measured with axial frequency $f_{\text{ip}} = 1.86\,\text{MHz}$, with Raman detuning $\Delta = 200\,\text{GHz}$ and before the reduction of phase chirps and radial mode coupling. Yet the achieved fidelities agree with predicted errors only due to out-of-Lamb-Dicke effects. This is because these sequences were optimised empirically using the Nelder-Mead algorithm. The optimisation lead to significant deviations from some of the theoretically-predicted parameter values. This suggests that effects of pulse-chirping can be largely compensated for by changes in the pulse parameters.

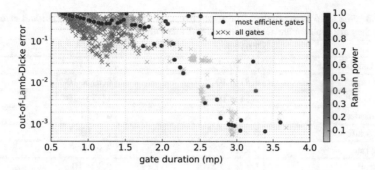

Fig. 7.13 Out-of-Lamb-Dicke errors: The plot shows the calculated error due to out-of-Lamb-Dicke effects of all simulated gate sequences. Most efficient gates are chosen as the sequence requiring the least Raman power in a given gate-time segment. The gates are colour-coded with the Raman power required for each sequence. Raman powers are normalised to the largest power required of any gate. The sequence that was implemented requiring the highest Raman power (0.89 mp, $P_{tot} \sim 300$ mW) corresponds to 0.07 on this scale. Sequences with simulated errors of <10% faster than 1.5 mp require at least three times that power

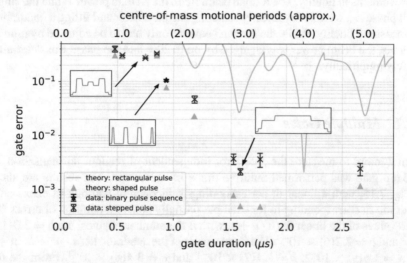

Fig. 7.14 Fast gate results: Shown are the measured fidelities for stepped and binary pulse sequences, together with their simulated errors. The theory curve for rectangular pulses is displayed for comparison. The insets show example pulse sequences. For gates slower than 1 μs the time in motional periods is slightly larger than indicated due to the increased axial frequency. The duration in μs is indicated accurately for all datapoints. The simulated error only includes the coherent error of the gate sequence, which is dominated by out-of-Lamb-Dicke effects. Figure published in [2]

Sequences with $t_g > 1$ μs were taken after the AOM RF frequencies were altered to minimise phase-chirps. The Raman beams had also been re-aligned to minimize coupling to radial modes. This also meant that the axial ip mode frequency had been changed to $f_z = 1.92$ MHz to maintain the correct ion spacing. The Raman detuning

Table 7.3 Fast Gate Error Budget The total error is a linear sum of the individual errors; this assumes they are constant and add incoherently. Table published in [2]

	3.05 mp	0.89 mp
Out-of-Lamb-Dicke	5×10^{-4}	3.1×10^{-1}
Phase chirp	$4(4) \times 10^{-4}$	$6(6) \times 10^{-3}$
Pulse control	$2(2) \times 10^{-4}$	$1(1) \times 10^{-3}$
Radial modes	4×10^{-5}	4×10^{-3}
Photon scattering	6×10^{-4}	7×10^{-3}
Heating rate	8×10^{-5}	3×10^{-5}
Total error	1.8×10^{-3}	3.3×10^{-1}

had been increased to $\Delta = -800\,\text{GHz}$ due to smaller power demands to reduce scattering errors. Pulse durations and amplitudes were at their optimum when set to their theoretical predictions and Nelder-Mead optimisation did not generate any improvements in fidelity. The Raman beatnote ν, the Raman power P and the single-qubit phase ϕ_{sq} were set by linearly scanning the parameter and fitting a quadratic to the measured fidelity. While the Raman beatnote only had to be adjusted by minimal amounts, the beam power was changed by up to 20% and the phase was determined entirely empirically.

7.4.1 Multiple Gates

In an attempt to measure the gate error independent of readout normalisation, the 3.05 mp gate was performed multiple times in succession. The results are shown in Fig. 7.15, with a linear fit corresponding to incoherent addition of errors and a quadratic fit corresponding to for example partially coherent addition of errors. The fitted values of the linear fit ($b \cdot n + c$, with n the number of gates) are $b = 3.3(1) \times 10^{-3}$ and $c = 2.2(6) \times 10^{-3}$. The fitted values of the quadratic fit ($a \cdot n^2 + b \cdot n + c$) are $a = 1.5(9) \times 10^{-4}$, $b = 2.1(7) \times 10^{-3}$ and $c = 3.4(8) \times 10^{-3}$. From the data collected no strong coherently adding error is visible. Deviations from the grey shaded area could either be explained by a small coherent error or by systematic differences between the single-gate and multiple-gate datasets. More data would be necessary to draw reliable conclusions.

7.4.2 Outlook

The gates implemented in this work are over an order of magnitude faster than the current highest-fidelity gates with trapped ions [4, 8], while still having fidelities

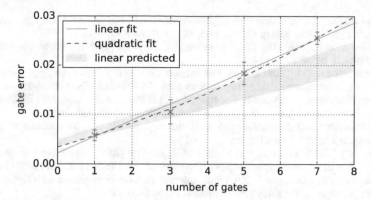

Fig. 7.15 Multiple gates: Gate errors measured for several 3.05 mp gates performed in succession. The gate errors are the raw measured errors not corrected for readout errors. The grey shaded area is the 1σ confidence interval for raw errors predicted by the readout error and fidelity measurement of the best single gate results, with $\epsilon_g = 2.2(3) \times 10^{-3}$

above the threshold for some quantum error correction schemes [9]. Although gates below $t_g = 1.5\,\mu$s exhibit large errors, the gate times achieved with high fidelities are of the same speed as typical single-qubit operations, and much shorter than most other elements in a measurement sequence, such as state preparation, cooling and readout. There is promising work on also reducing the speed of those elements, such as EIT cooling [10] for faster cooling to the ground state ($t = 80\,\mu$s for cooling both modes of a ^{9}Be$^+$–^{24}Mg$^+$ crystal to $\bar{n} \lesssim 0.1$), or direct fluorescence detection [11, 12] for readout within 10–50 μs.

To further reduce the speed of our two-qubit entanglement gates there are several options. By increasing the Raman intensity at the ions less power-efficient but faster gates can be implemented. Equally, reducing the angle between the two gate Raman-beams and therefore reducing the Lamb-Dicke parameter will shift the speed-limit when out-of-Lamb-Dicke effects start to matter. However a reduced Lamb-Dicke parameter also means that larger Raman powers are necessary, leading to larger scattering errors. Finally, the active stabilisation of the relative Raman phase would drastically reduce errors due to out-of-Lamb-Dicke effects and allow the implementation of even faster gates. Coherent errors of the gate sequence scale with the number of motional periods rather than the absolute gate duration. Using a lighter ion such as ^{9}Be$^+$ allows larger axial frequencies f_z and would roughly halve the gate duration.

References

1. Ballance CJ (2014) High-fidelity quantum logic in Ca$^+$ PhD thesis, University of Oxford
2. Schäfer VM et al (2017) Fast quantum logic gates with trapped-ion qubits. Nature 555:75–78. ISSN: 0028-0836

3. Leibfried D et al (2003) Experimental demonstration of a robust, high-fidelity geometric two ion-qubit phase gate. Nature 422:412–415. ISSN: 0028-0836
4. Gaebler JP et al (2016) High-fidelity universal gate set for $^9Be^+$ ion qubits. Phys Rev Lett 117:060505. ISSN: 10797114
5. Degenhardt C et al (2005) Influence of chirped excitation pulses in an optical clock with ultracold calcium atoms. IEEE Trans Instrum Meas 54:771–775. ISSN: 00189456
6. Steane AM, Imreh G, Home JP Leibfried D (2014) Pulsed force sequences for fast phase-insensitive quantum gates in trapped ions. New J Phys 16. ISSN: 13672630
7. Woodrow SR (2015) Linear Paul trap design for high-fidelity, scalable quantum information processing Master's thesis, University of Oxford
8. Ballance CJ, Harty TP, Linke NM, Sepiol MA, Lucas DM (2016) High-fidelity quantum logic gates using trapped-ion hyperfine qubits. Phys Rev Lett 117:060504. ISSN: 10797114
9. Raussendorf R, Harrington J (2007) Fault-tolerant quantum computation with high threshold in two dimensions. Phys Rev Lett 98:190504. ISSN: 00319007
10. Lin Y et al (2013) Sympathetic electromagnetically-induced-transparency laser cooling of motional modes in an ion chain. Phys Rev Lett 110:153002. ISSN: 00319007
11. Noek R et al (2013) High speed, high fidelity detection of an atomic hyperfine qubit. Opt Lett 38:4735–4738. ISSN: 0146-9592
12. Hughes A (2017) A new readout method for $^{43}Ca^+$ qubits tech. rep. August, University of Oxford

Chapter 8
Conclusion

We have demonstrated two-qubit entangling gates between ions of different species, using a technique that requires only a single laser for both species. We first demonstrated the technique with two isotopes of calcium, ^{40}Ca$^+$ and ^{43}Ca$^+$, and also performed a Bell test with these non-identical particles. We then used the same gate mechanism to perform an entangling gate on two different atomic elements, ^{43}Ca$^+$ and ^{88}Sr$^+$, in a proof-of-principle experiment. Entangling these different species of ion is an important step towards an approach for building a scalable quantum computer with trapped ions, by using ion-photon entanglement to connect ions in different trap modules together.

In a separate experiment we have performed fast entangling gates with trapped ions. These gates are outside of the adiabatic regime and therefore of similar speed to the motion of the ions in the trap. For this purpose we have shaped the amplitude of the laser pulses driving the gate. The pulse sequences were designed such that the gate performance is independent of the phase of the optical driving field—a parameter that is difficult to control in the lab. With this technique we decreased the duration of two-qubit gates by over an order of magnitude, without any considerable increase in the gate errors. We also demonstrated entanglement creation within less than a single motional period of the ions; however at these gate speeds out-of-Lamb-Dicke effects caused considerable errors.

8.1 Comparison with Other Trapped-Ion Results

8.1.1 Mixed-Species Entanglement

Sympathetic cooling of different species of ion was first performed in 1980 (1986) with different isotopes of Mg$^+$ (different atomic species ^{9}Be$^+$ and ^{198}Hg$^+$). The first demonstration of quantum logic spectroscopy was in 2005, using a ^{9}Be$^+$ ion as logic ion and an ^{27}Al$^+$ ion as the spectroscopy ion [1]. The work reported here and

© Springer Nature Switzerland AG 2020
V. M. Schäfer, *Fast Gates and Mixed-Species Entanglement with Trapped Ions*,
Springer Theses, https://doi.org/10.1007/978-3-030-40285-3_8

published in [2] was the first demonstration of an entangling gate between different species of ion—together with [3] who performed a two-qubit gate with Bell state fidelity $\mathscr{F} = 97.9(1)\%$ between $^9\text{Be}^+$ and $^{25}\text{Mg}^+$. In 2017 mixed-species entanglement with $\mathscr{F} = 60\%$ was reported with ion species $^{171}\text{Yb}^+$ and $^{138}\text{Ba}^+$[4].

8.1.2 Fast Gates

The fidelity of trapped-ion two-qubit gates has been increasing steadily, with fidelities $\mathscr{F} > 99\%$ achieved in 2008 [5] and $\mathscr{F} \approx 99.9\%$ in 2016 [6, 7]. The gate durations however have remained fairly constant, being limited by the motional frequencies of the ions in the trap. Faster gate times can be achieved with lighter ions such as $^9\text{Be}^+$, that attain higher trap frequencies for a given trap voltage [7, 8], and errors due to off-resonant excitation can be reduced using pulse-shaping techniques for moderately fast gate times [9]. However, when the gate time approaches a single motional period of the ions, gate errors rapidly increase with the conventional gate mechanisms. A list of published trapped-ion two-qubit gates with their gate durations and fidelities is shown in Table 8.1.

The first fast gate mechanism breaking this speed limit was proposed in 2003 [16] and uses short laser pulses from different directions, interspersed with periods of free evolution. Because the short laser pulses do not couple to specific modes of motion, but instead give rise to the gate dynamics via mechanical kicks to the ion's momentum, this gate mechanism is insensitive to the ions' temperature and ooLD effects. The scheme has been expanded to allow entangling gates on neighbouring ions in

Table 8.1 Trapped-ion two-qubit gates: The gate times quoted are from the beginning of the rising edge to the end of the falling edge, and were derived from the gate durations and pulse shaping times, where accessible

Gate time (μs)	Fidelity (%)	Qubit	Year	Author
39	97(2)	$^9\text{Be}^+$ hyperfine	2003	Leibfried et al. [10]
52	99.3(1)	$^{40}\text{Ca}^+$ optical	2008	Benhelm et al. [5]
27.5	97.1(2)	$^{40}\text{Ca}^+$ optical	2009	Kirchmair et al. [11]
37.5	98(2)	$^{171}\text{Yb}^+$ hyperfine	2009	Kim et al. [12]
20	91(2)	$^9\text{Be}^+$ hyperfine	2012	Gaebler et al. [8]
105	94.6(4)	$^9\text{Be}^+$ hyperfine	2013	Tan et al. [13]
130	98.1(6)	$^{88}\text{Sr}^+$ optical	2014	Navon et al. [14]
100	99.89(7)	$^{43}\text{Ca}^+$ hyperfine	2016	Ballance et al. [6]
5.3	97.1(2)	$^{43}\text{Ca}^+$ hyperfine	2016	Ballance et al. [6]
30.8	99.92(4)	$^9\text{Be}^+$ hyperfine	2016	Gaebler et al. [7]
18.5	76(1)	$^{171}\text{Yb}^+$ hyperfine	2017	Wong-Campos et al. [15]

a long string of ions [17], without being sensitive to the spectral mode bunching. This provides a means of scaling up trapped-ion systems without the need for (or with fewer) shuttling operations than are necessary in the QCCD approach. However sufficiently precise control of the fast laser pulses turned out to be difficult.

The first experimental demonstration using fast momentum kicks was in 2013, when the scheme was used to create entanglement between the spin and motion of a single ion [18]. Subsequently multiple concatenated kicks were used to create large Schrödinger cat states of a single ion [19]. Simultaneously with our results the implementation of a two-qubit gate using spin dependent kicks was reported [15]. While the scheme can in principle achieve gate speeds below a single motional period of an ion, the gate implemented in [15] had a longer gate duration limited by technical constraints of $t_{gt} = 18.5\,\mu s$, about 23 motional periods, with an achieved gate fidelity $\mathscr{F} = 76(1)\%$.

Recently another fast gate mechanism has been proposed [20], which also uses a variation of the σ_z geometric phase gate with amplitude-shaped pulses. It obtains analytic solutions for smooth forces, at the price of sensitivity to the relative Raman phase ϕ_0. As for our gate-mechanism the scheme is sensitive to ooLD effects.

The work described in this thesis and published in [21] is the first demonstration of entanglement generation below a single motional period of the ions. The $1.6\,\mu s$ gate is faster than previous trapped-ion gates above the error-correction threshold by a factor of 20–60, without considerable increase in gate errors. A summary of the fastest and highest fidelity trapped-ion two-qubit gates is shown in Fig. 8.1.

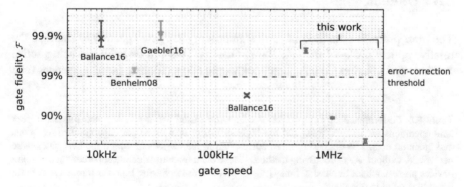

Fig. 8.1 Trapped ion two-qubit gates versus gate speed. The plotted gates are selected for high fidelities and fast gate speeds. The displayed error correction threshold holds for some surface code error correction schemes [22]. However, for feasible implementation of error correction without the need for an overly large overhead of ancilla qubits, gate errors considerably below this threshold are desirable. A more exhaustive list of gates can be found in Table 8.1

8.2 Comparison with Other Qubit Platforms

A comparison between gate durations, gate fidelities and coherence times for different qubit platforms is shown in Table 8.2. While trapped ions have significantly better coherence times and gate fidelities, the duration of their two-qubit and typical single-qubit operations is considerably longer than those of superconducting qubits and quantum dots. With specialised gate mechanisms significantly faster gates are possible, even exceeding the speed of solid state single-qubit gates. For all platforms trapped ions have the highest ratio of coherence time to gate duration, an important benchmark for a quantum computer that needs to implement many gate operations in succession.

The fast gates in this work were implemented on the stretch qubit which has lower coherence time, because it is based on the σ_z geometric phase gate that can not be performed on clock qubits. It is not trivial to use the same technique for fast gates with a Mølmer-Sørensen gate [30], that could be performed on clock qubits. However the coherence time of the stretch qubit is only limited by magnetic field noise, which can be reduced with technical improvements. Current efforts include the usage of superconducting coils for more stable magnetic fields and better shielding. Our gate was performed at a magnetic field giving access to a clock qubit in ^{43}Ca$^+$, and the qubit states could therefore be mapped onto the clock qubit between operations.

8.3 Outlook

The most promising method of further improving the gate speed is to stabilise the relative optical phase of the gate Raman beams. This can be achieved using active feedback from the ion [31]. A phase lock would allow tailoring pulse sequences for a

Table 8.2 Performance of different kinds of qubits: Listed are the fastest and highest-fidelity gate operations for trapped ions, and the highest-fidelity operations for superconducting qubits and quantum dots. The dephasing times correspond to the devices in which the best gates were performed, without any re-focussing methods. Some other quantum dot and superconducting qubit devices possess longer dephasing times [28, 29]. The 5 ms dephasing time for trapped ions is for the stretch qubit in this work

Qubit type	Dephasing time	Single-qubit gate		Two-qubit gate	
		Error	Duration	Error	Duration
Trapped ions	$T_2^* = 50$ s [23] $T_2^* = 1.5$ s [7] $T_2^* = 5(1)$ ms [21]	$1.0(3) \times 10^{-6}$ [23] 7×10^{-3} [24]	$12\,\mu$s [23] $40\,$ps [24]	$8(4) \times 10^{-4}$ [7] $2.2(3) \times 10^{-3}$ [21]	$30\,\mu$s [7] $1.6\,\mu$s [21]
Superconducting qubits	$T_2 \sim 10\,\mu$s [25]	$6.0(5) \times 10^{-4}$ [25]	$15\,$ns [25]	$5.6(5) \times 10^{-3}$ [25]	$43\,$ns [25]
Quantum dots	$T_2^* = 20.4\,\mu$s [26] $T_2^* = 0.8\,\mu$s [27]	$7.4(2) \times 10^{-4}$ [26]	$120\,$ns [26]	$1.1(2) \times 10^{-1}$ [27]	$\sim 130\,$ns [27]

specific initial phase ϕ_0 and thus reduce errors due to ooLD effects. In addition, other gate schemes such as the proposal by Palmero et al. [20] could be implemented and compared. Performing fast gates on clock qubits with greatly increased coherence times would be another important step, but requires further theoretical work.

Following our work on entangling ^{88}Sr$^+$ and ^{43}Ca$^+$, we are currently building a system of two identical ion traps connected via a photonic link. With only 5 ions in each trap and realistic models for gate fidelities and noise in the optical link, entanglement purification protocols can be implemented [32]. This would make such a module a scalable building block for a quantum computer. The vacuum systems are equipped with a Sandia HOA2 multi-zone surface trap, which promises better scalability than the three dimensional macroscopic blade trap used in this work.

For a realistic prospect of building a quantum computer with trapped ions there are still several tasks to be achieved. While the errors of current two-qubit entangling gates are below the 'threshold' for quantum error correction schemes, the overhead of physical qubits necessary to encode a logical qubit is still daunting. Further reducing these errors should therefore be a priority. The added complexity of experiments comprising several traps and using multiple species makes it necessary to improve stability and miniaturisation of laser and control setups. Technological advances in fibre-integrated optics and AOMs can greatly reduce experimental complexity. Progress in optics integrated into trap chips not only aids miniaturisation but also reduces power requirements and promises better beam pointing stability [33]. For using a design of photonically-linked trap modules it is important to improve ion-photon coupling efficiency beyond what is possible with a lens in free space. Optical cavities integrated into the ion trap promise greatly improved coupling efficiencies, but still require substantial experimental progress. Advances in surface traps, two-qubit entangling gates using microwaves [34] and microwave addressing [35] allow the use of this more mature and scalable technology. In addition it gives the prospect of reducing gate errors below the limit of photon scattering inherent to laser gates.

While there remain plenty of tasks to solve, progress in recent years has been steady. With the prospect of actually building a quantum computer in sight, these are exciting times to work on trapped ion quantum computing.

References

1. Schmidt PO et al (2005) Spectroscopy using quantum logic. Sci 309:749–752. ISSN: 0036-8075
2. Ballance CJ et al (2015) Hybrid quantum logic and a test of Bell's inequality using two different atomic isotopes. Nature 528:384–386. ISSN: 0028-0836
3. Tan TR et al (2015) Multi-element logic gates for trapped-ion qubits. Nature 528:380–383. ISSN: 0028-0836
4. Inlek IV, Crocker C, Lichtman M, Sosnova K, Monroe C (2017) Multispecies trapped-ion node for quantum networking. Phys Rev Lett 118:250502
5. Benhelm J, Kirchmair G, Roos CF, Blatt R (2008) Towards fault-tolerant quantum computing with trapped ions. Nat Phys 4:463–466. ISSN: 1745-2473

6. Ballance CJ, Harty TP, Linke NM, Sepiol MA, Lucas DM (2016) High-fidelity quantum logic gates using trapped-ion hyperfine qubits. Phys Rev Lett 117:060504. ISSN: 10797114
7. Gaebler JP et al (2016) High-fidelity universal gate set for 9Be+ ion qubits. Phys Rev Lett 117:060505. ISSN: 10797114
8. Gaebler JP et al (2012) Randomized benchmarking of multiqubit gates. Phys Rev Lett 108:260503. ISSN: 0031-9007
9. Ballance CJ (2014) High-fidelity quantum logic in Ca + PhD thesis (University of Oxford, 2014)
10. Leibfried D et al (2003) Experimental demonstration of a robust, high-fidelity geometric two ion-qubit phase gate. Nature 422:412–415. ISSN: 0028-0836
11. Kirchmair G et al (2009) Deterministic entanglement of ions in thermal states of motion. New J Phys 11. ISSN: 13672630
12. Kim K et al (2009) Entanglement and tunable spin-spin couplings between trapped ions using multiple transverse modes. Phys Rev Lett 103:120502. ISSN: 00319007
13. Tan TR et al (2013) Demonstration of a dressed-state phase gate for trapped ions. Phys Rev Lett 110:263002. ISSN: 00319007
14. Navon N, Akerman N, Kotler S, Glickman Y, Ozeri R (2014) Quantum process tomography of a Mølmer-Sørensen interaction. Phys Rev A 90:010103. ISSN: 10941622
15. Wong-Campos JD, Moses SA, Johnson KG, Monroe C (2017) Demonstration of two-atom entanglement with ultrafast optical pulses. Phys Rev Lett 119:230501. ISSN: 0031-9007
16. García-Ripoll JJ, Zoller P, Cirac JI (2003) Speed optimized two-qubit gates with laser coherent control techniques for ion trap quantum computing. Phys Rev Lett 91:157901. ISSN: 0031-9007
17. Duan L-M (2004) Scaling ion trap quantum computation through fast quantum gates. Phys Rev Lett 93:100502
18. Mizrahi J et al (2013) Ultrafast spin-motion entanglement and interferometry with a single atom. Phys Rev Lett 110:203001. ISSN: 00319007
19. Johnson KG, Wong-Campos JD, Neyenhuis B, Mizrahi J, Monroe C (2017) Ultrafast creation of large Schrödinger cat states of an atom. Nature Commun 8. ISSN: 20411723
20. Palmero M, Martinez-Garaot S, Leibfried D, Wineland DJ, Muga JG (2017) Fast phase gates with trapped ions. Phys Rev A 95:022328
21. Schäfer VM et al (2017) Fast quantum logic gates with trapped-ion qubits. Nature 555:75–78. ISSN: 0028-0836
22. Fowler AG, Mariantoni M, Martinis JM, Cleland AN (2012) Surface codes: towards practical large-scale quantum computation. Phys Rev A 86:032324. ISSN: 10502947
23. Harty TP et al (2014) High-fidelity preparation, gates, memory, and readout of a trapped-ion quantum bit. Phys Rev Lett 113:220501. ISSN: 0031-9007
24. Campbell WC et al (2010) Ultrafast gates for single atomic qubits. Phys Rev Lett 105:090502. ISSN: 00319007
25. Barends R et al (2014) Superconducting quantum circuits at the surface code threshold for fault tolerance. Nature 508:500–503. ISSN: 0028-0836
26. Yoneda J et al, A quantum-dot spin qubit with coherence limited by charge noise and fidelity higher than 99.9
27. Watson TF et al (2018) A programmable two-qubit quantum processor in silicon. Nature 555, 633–637. ISSN: 0028-0836
28. Veldhorst M et al (2014) An addressable quantum dot qubit with fault-tolerant control-fidelity. Nature Nanotechnol 9:981–985. ISSN: 17483395
29. Wendin G (2017) Quantum information processing with superconducting circuits: a review. Reports Progress Phys 80:106001. ISSN: 0034-4885 (2017)
30. Steane AM, Imreh G, Home JP, Leibfried D (2014) Pulsed force sequences for fast phase-insensitive quantum gates in trapped ions. New J Phys 16. ISSN: 13672630
31. Schmiegelow CT et al (2016) Phase-stable free-space optical lattices for trapped ions. Phys Rev Lett 116:033002. ISSN: 10797114

32. Nigmatullin R, Ballance CJ, de Beaudrap N, Benjamin SC (2016) Minimally complex ion traps as modules for quantum communication and computing. New J Phys 18:103028. ISSN: 13672630
33. Mehta KK et al (2016) Integrated optical addressing of an ion qubit. Nature Nanotechnol 11:1066–1070. ISSN: 17483395
34. Harty TP et al (2016) High-fidelity trapped-ion quantum logic using near-field microwaves. Phys Rev Lett 117: 140501. ISSN: 10797114
35. Aude Craik DPL et al (2017) High-fidelity spatial and polarization addressing of 43Ca+ qubits using near-field microwave control. Phys Rev A 95:022337. ISSN: 24699934

Appendix A
Useful Identities and Definitions

A.1 Spin Operators

The Pauli matrices are:

$$\sigma_z = \begin{pmatrix} 1 & 0 \\ 0 & -1 \end{pmatrix}, \ \sigma_y = \begin{pmatrix} 0 & -i \\ i & 0 \end{pmatrix}, \ \sigma_x = \begin{pmatrix} 0 & 1 \\ 1 & 0 \end{pmatrix} \tag{A.1}$$

with excited state $|e\rangle = \begin{pmatrix} 1 \\ 0 \end{pmatrix}$, ground state $|g\rangle = \begin{pmatrix} 0 \\ 1 \end{pmatrix}$ and $\sigma_z|e\rangle = 1, \sigma_z|g\rangle = -1$. Useful identities are:

$$\sigma_z = |e\rangle\langle e| - |g\rangle\langle g|, \ \sigma_y = i\left(|g\rangle\langle e| - |e\rangle\langle g|\right), \ \sigma_x = |e\rangle\langle g| + |g\rangle\langle e| \tag{A.2}$$

and

$$\sigma_+ = \frac{1}{2}(\sigma_x + i\sigma_y) = \begin{pmatrix} 0 & 1 \\ 0 & 0 \end{pmatrix} \tag{A.3}$$

$$\sigma_- = \frac{1}{2}(\sigma_x - i\sigma_y) = \begin{pmatrix} 0 & 0 \\ 1 & 0 \end{pmatrix}, \tag{A.4}$$

The commutation relations are

$$[\sigma_x, \sigma_y] = 2i\epsilon_{xyz}\sigma_z \tag{A.5}$$

$$[\sigma_+, \sigma_-] = \sigma_z \tag{A.6}$$

$$[\sigma_+, \sigma_z] = -2\sigma_+ \tag{A.7}$$

$$[\sigma_-, \sigma_z] = 2\sigma_- \tag{A.8}$$

© Springer Nature Switzerland AG 2020
V. M. Schäfer, *Fast Gates and Mixed-Species Entanglement with Trapped Ions*,
Springer Theses, https://doi.org/10.1007/978-3-030-40285-3

For calculating phase propagators:

$$R_z(\phi) = e^{i\phi\sigma_z} = \begin{pmatrix} 1 & \\ & e^{-i2\phi} \end{pmatrix} \tag{A.9}$$

$$e^{i\phi(\sigma_z \otimes \sigma_z)} = \begin{pmatrix} 1 & & & \\ & e^{-i2\phi} & & \\ & & e^{-i2\phi} & \\ & & & 1 \end{pmatrix} \tag{A.10}$$

Single qubit rotation operators around x- and y-axis:

$$R_x(\theta) = \begin{pmatrix} \cos(\theta/2) & -i\sin(\theta/2) \\ -i\sin(\theta/2) & \cos(\theta/2) \end{pmatrix} \tag{A.11}$$

$$R_y(\theta) = \begin{pmatrix} \cos(\theta/2) & \sin(\theta/2) \\ -\sin(\theta/2) & \cos(\theta/2) \end{pmatrix} \tag{A.12}$$

Trigonometric identities:

$$\sin(x) = \sum_{n=1}^{\infty} \frac{(-1)^{n-1}}{(2n-1)!} x^{2n-1} \tag{A.13}$$

$$\cos(x) = \sum_{n=0}^{\infty} \frac{(-1)^n}{(2n)!} x^{2n} \tag{A.14}$$

$$\exp(X) = \sum_{n=0}^{\infty} \frac{1}{n!} X^n \tag{A.15}$$

A.2 Motion Operators

Ladder operators:

$$a^\dagger |n\rangle = \sqrt{n+1} \, |n+1\rangle \tag{A.16}$$

$$a |n\rangle = \sqrt{n} \, |n-1\rangle \tag{A.17}$$

$$a^\dagger a = N \tag{A.18}$$

$$N |n\rangle = n |n\rangle \tag{A.19}$$

In matrix form:

$$a^\dagger = \begin{pmatrix} 0 & 0 & 0 & 0 & \dots \\ \sqrt{1} & 0 & 0 & 0 & \dots \\ 0 & \sqrt{2} & 0 & 0 & \dots \\ 0 & 0 & \sqrt{3} & 0 & \dots \\ \vdots & \vdots & \vdots & & \ddots \end{pmatrix} \tag{A.20}$$

$$a = \begin{pmatrix} 0 & \sqrt{1} & 0 & 0 & \dots \\ 0 & 0 & \sqrt{2} & 0 & \dots \\ 0 & 0 & 0 & \sqrt{3} & \dots \\ 0 & 0 & 0 & 0 & \ddots \\ \vdots & \vdots & \vdots & \vdots & \ddots \end{pmatrix} \tag{A.21}$$

The commutation relations are:

$$[a, a^\dagger] = 1 \tag{A.22}$$

$$[N, a^\dagger] = a^\dagger \tag{A.23}$$

$$[N, a] = -a \tag{A.24}$$

Thermal states:

$$\rho = \frac{1}{Z} \sum_{n=0}^{\infty} e^{-n\hbar\beta\omega_z} |\alpha, n\rangle\langle\alpha, n| \tag{A.25}$$

$$= \sum_{n=0}^{\infty} \frac{\bar{n}^n}{(\bar{n}+1)^{n+1}} |\alpha, n\rangle\langle\alpha, n| \tag{A.26}$$

where $Z = \frac{1}{1-e^{-\hbar\beta\omega_z}} = \bar{n} + 1, \bar{n} = <n>_{\text{th}}, \beta = 1/(k_B T)$

Generalized Laguerre polynomials:

$$\mathscr{L}_n^{(\alpha)}(x) = \sum_{m=0}^{n} \binom{n+\alpha}{n-m} \frac{(-x)^m}{m!} \tag{A.27}$$

Appendix B
Matrix Elements

For arbitrary polarisations $\hat{b} = (b_+\hat{\sigma}^+ + b_0\hat{\pi} + b_-\hat{\sigma}^-)$ and $\hat{r} = (r_+\hat{\sigma}^+ + r_0\hat{\pi} + r_-\hat{\sigma}^-)$.

B.1 Calcium 43

$$\Omega_R = \frac{\sqrt{7}\omega_f \left(b_- r_0 + b_0 r_+\right)}{12\Delta \left(\Delta - \omega_f\right)}$$

$$\Omega_\downarrow = \frac{3\Delta \left(b_+ r_+ + b_- r_- + b_0 r_0\right) - \omega_f \left(2b_- r_- + b_0 r_0\right)}{6\Delta \left(\Delta - \omega_f\right)}$$

$$\Omega_\uparrow = \frac{12\Delta \left(b_0 r_0 + b_+ r_+ + b_- r_-\right) - \omega_f \left(7b_+ r_+ + b_- r_- + 4b_0 r_0\right)}{24\Delta \left(\Delta - \omega_f\right)}$$

$$\Delta_{LS} = -\frac{7\omega_f \left(e_+^2 - e_-^2\right)}{48\Delta \left(\Delta - \omega_f\right)}$$

$$\Gamma_{tot} = \frac{24\Delta^2 + (\omega_f^2 - 2\omega_f\Delta)(7 + 2e_-^2 + e_0^2)}{96\Delta^2(\Delta - \omega_f)^2}$$

$$\Gamma_{Raman,\uparrow} = \frac{\omega_f^2 \left(35e_+^2 + 11e_-^2 + 32e_0^2\right)}{576\Delta^2 \left(\Delta - \omega_f\right)^2}$$

$$\Gamma_{Raman,\downarrow} = \frac{\omega_f^2 \left(e_-^2 + e_0^2\right)}{18\Delta^2 \left(\Delta - \omega_f\right)^2}$$

$$\Gamma_{el} = \frac{49\omega_f^2 \left(e_+^2 + e_-^2\right)}{576\Delta^2 \left(\Delta - \omega_f\right)^2}$$

© Springer Nature Switzerland AG 2020
V. M. Schäfer, *Fast Gates and Mixed-Species Entanglement with Trapped Ions*,
Springer Theses, https://doi.org/10.1007/978-3-030-40285-3

B.2 Strontium 88

$$\Omega_R = -\frac{\sqrt{2}\omega_f\,(b_+r_0 + b_0r_-)}{6\Delta\,(\Delta - \omega_f)}$$

$$\Omega_\uparrow = \frac{3\Delta\,(b_+r_+ + b_-r_- + b_0r_0) - \omega_f\,(2b_-r_- + b_0r_0)}{6\Delta\,(\Delta - \omega_f)}$$

$$\Omega_\downarrow = \frac{3\Delta\,(b_+r_+ + b_-r_- + b_0r_0) - \omega_f\,(2b_+r_+ + b_0r_0)}{6\Delta\,(\Delta - \omega_f)}$$

$$\Delta_{LS} = \frac{\omega_f\,(e_+^2 - e_-^2)}{6\Delta\,(\Delta - \omega_f)}$$

$$\Gamma_{Raman,\downarrow} = \frac{\omega_f^2\,(e_+^2 + e_0^2)}{18\Delta^2\,(\Delta - \omega_f)^2}$$

$$\Gamma_{Raman,\uparrow} = \frac{\omega_f^2\,(e_-^2 + e_0^2)}{18\Delta^2\,(\Delta - \omega_f)^2}$$

$$\Gamma_{tot} = \frac{9\Delta^2 + 2\omega_f^2 + (3\Delta - 2\omega_f)^2}{72\Delta^2\,(\Delta - \omega_f)^2}$$

$$\Gamma_{el} = \frac{\omega_f^2\,(e_+^2 + e_-^2)}{9\Delta^2\,(\Delta - \omega_f)^2}$$

Appendix C
Two Qubit Gates

C.1 Magnus Expansion

Calculating the propagator of a time-dependent Hamiltonian can be tricky, because the Hamiltonian at different times do not commute with each other. Therefore, instead of $U = e^{-iHt/\hbar}$ the Dyson time ordering operator $U = \mathscr{T}\left(e^{-\frac{i}{\hbar}\int_0^t H(t')dt'}\right)$ has to be used. It can often be favourable to use the Magnus expansion [1, 2] instead, with

$$U(t) = e^{\Omega(t)} \tag{C.1}$$

$$\Omega(t) = \sum_{k=1}^{\infty} \Omega_k(t)$$

$$= \int_0^t A(\tau)d\tau$$

$$+ \frac{1}{2}\int_0^t d\tau \int_0^\tau d\sigma \, [A(\tau), A(\sigma)]$$

$$+ \frac{1}{6}\int_0^t d\tau \int_0^\tau d\sigma \int_0^\sigma d\rho \, ([A(\tau), [A(\sigma), A(\rho)]] + [A(\rho), [A(\sigma), A(\tau)]])$$

$$+ \dots$$

for

$$\frac{dU(t)}{dt} = A(t)U(t), \quad U(0) = I \tag{C.2}$$

We can identify U as the propagator in the Schrödinger picture $\psi(t) = U(t)\psi(0)$ from the Schrödinger equation $H\psi(t) = i\hbar\frac{\partial}{\partial t}\psi(t)$, with $A = -\frac{i}{\hbar}H(t)$. One advantage of the Magnus expansion over the Dyson series is that the unitarity of the propagator U is preserved for all orders of approximation. However this approach is especially favourable for calculating the propagator of geometric phase gates, where

© Springer Nature Switzerland AG 2020

V. M. Schäfer, *Fast Gates and Mixed-Species Entanglement with Trapped Ions*,
Springer Theses, https://doi.org/10.1007/978-3-030-40285-3

the Hamiltonian has the shape $H \sim \gamma a + \gamma^* a^\dagger$. Here the commutator $[A(\tau), A(\sigma)]$ is scalar, causing all higher orders to vanish and the resulting propagator to be an exact solution.

C.1.1 Geometric Phase Gate

The Hamiltonian for the geometric phase gate is

$$\hat{H}_{eff}(t) = \sum_{m=\uparrow,\downarrow} \frac{\hbar\Omega_m}{2} i\eta \, |m\rangle\langle m| \left[a e^{i(\delta_g t - \phi_0)} - a^\dagger e^{-i(\delta_g t - \phi_0)} \right] \tag{C.3}$$

from which we obtain

$$A = \dot{\gamma}(t) a - \dot{\gamma}^*(t) a^\dagger \tag{C.4}$$

where $\dot{\gamma}(t) = \sum_{m=\uparrow,\downarrow} \frac{\Omega_m \eta}{2} e^{i(\delta_g t - \phi_0)} |m\rangle\langle m| = \beta e^{i\delta_g t}$. From this we can calculate

$$\Omega_1 = \left([\gamma(t) - \gamma(0)] a - [\gamma^*(t) - \gamma^*(0)] a^\dagger \right) |m\rangle\langle m|$$
$$= \sum_{m=\uparrow,\downarrow} \left(\alpha_m a^\dagger - \alpha_m^* a \right) |m\rangle\langle m| \tag{C.5}$$

with $\alpha_m = -\frac{\Omega_m \eta}{\delta_g} e^{-i\phi_0} e^{-i\delta_g t/2} \sin\left(\frac{\delta_g t}{2}\right)$ and

$$\Omega_2 = -i \frac{|\beta|^2}{\delta_g^2} [t\delta_g - \sin(\delta_g t)]$$
$$= -i\Phi \tag{C.6}$$

with $\Phi = \frac{\eta^2}{4\delta_g^2} [\delta_g t - \sin(\delta_g t)] \left(\sum_{m=\uparrow,\downarrow} \Omega_m |m\rangle\langle m| \right)^2$. This yields the exact propagator

$$U(t) = \sum_{m=\uparrow,\downarrow} D(\alpha_m |m\rangle\langle m|) e^{-i\Phi} \tag{C.7}$$

with displacement operator $D(\alpha) = e^{\alpha a^\dagger - \alpha^* a}$.

Table C.1 Geometric phase gate Rabi frequencies: Rabi frequencies for a geometric phase-gate with two ions with unequal force amplitudes on $|\uparrow\rangle$ and $|\downarrow\rangle$. The asymmetry of the matrix elements leads to an additional single-qubit phase $I \otimes \begin{pmatrix} 1 & 0 \\ 0 & e^{i\phi_a} \end{pmatrix} + \begin{pmatrix} 1 & 0 \\ 0 & e^{i\phi_a} \end{pmatrix} \otimes I$

Spin state	$e^{i\phi_z} = 1$		$e^{i\phi_z} = -1$	
	ip	oop	ip	oop
$\uparrow\uparrow$	$2\Omega_\uparrow$	0	0	$2\Omega_\uparrow$
$\downarrow\downarrow$	$-2\Omega_\uparrow - 2\delta\Omega$	0	0	$-2\Omega_\uparrow - 2\delta\Omega$
$\uparrow\downarrow$	$-\delta\Omega$	$2\Omega_\uparrow + \delta\Omega$	$2\Omega_\uparrow + \delta\Omega$	$-\delta\Omega$
$\downarrow\uparrow$	$-\delta\Omega$	$-2\Omega_\uparrow - \delta\Omega$	$-2\Omega_\uparrow - \delta\Omega$	$-\delta\Omega$

C.2   Unequal Force Amplitudes

For $\Omega_\downarrow = -\Omega_\uparrow - \delta\Omega$ we obtain for integer ion spacings ($e^{i\phi_z} = 1$, $\delta z = k\lambda_z$) and half-integer ion spacings ($e^{i\phi_z} = -1$, $\delta z = (k + \frac{1}{2})\lambda_z$) respectively the effective Rabi-frequency $\Omega = \Omega_{s_1} + \text{sign}(m)e^{i\phi_z}\Omega_{s_2}$, where $m = 1$ for the in-phase mode and $m = -1$ for the out-of-phase mode. The resulting Rabi-frequencies are listed in Table C.1.

C.3   Fast Gate Thermal Error

We can calculate the thermal part of the fast gates error

$$\epsilon_{th} = 1 - \sum_{m,s} \frac{1}{4}\Big\langle \left|\langle \Delta\alpha_m^s, n_m | 0, n_m \rangle\right|^2 \Big\rangle_{th} \tag{C.8}$$

using the identities [3]

$$\langle \beta, m | \alpha, n \rangle = \langle \beta | \alpha \rangle \sqrt{\frac{m!}{n!}} (\beta^* - \alpha^*)^{n-m} \mathscr{L}_m^{n-m}\left(|\beta - \alpha|^2\right) \tag{C.9}$$

$$\langle \beta | \alpha \rangle = \exp\left\{\alpha\beta^* - \frac{1}{2}\left(|\alpha|^2 + |\beta|^2\right)\right\} \tag{C.10}$$

and the sum identity for the squared generalised Laguerre polynomial with $|z| < 1$ [4]

$$\sum_{n=0}^{\infty} z^n \mathscr{L}_n(x)^2 = \frac{e^{-2xz/(1-z)}}{1-z} \mathscr{J}_0\left(\frac{2x\sqrt{z}}{1-z}\right) \tag{C.11}$$

The 0th-order Bessel function is $\mathscr{J}_0(\epsilon) = 1 + \epsilon^2/4 + \dots$. Expanding to lowest order we obtain

$$\epsilon_{\text{th}} = 1 - \sum_{m,s} \frac{1}{4} e^{-|\Delta\alpha_m^s|^2} \sum_{n=0}^{\infty} \frac{\bar{n}^n}{(\bar{n}+1)^{n+1}} \left| \mathscr{L}_n^0 \left(|\Delta\alpha_m^s|^2 \right) \right|^2 \qquad \text{(C.12)}$$

$$\approx \sum_{m,s} \frac{1+2\bar{n}}{4} |\Delta\alpha_m^s|^2 \qquad \text{(C.13)}$$

C.4   Phases

Understanding and controlling phases is a key requirement to performing high-fidelity two-qubit gates. We therefore briefly summarize all important phases, their origin and how to control them. We choose as basis for the space of phases

$$\hat{\theta}_1 = \begin{pmatrix} 1 & & & \\ & 1 & & \\ & & -1 & \\ & & & -1 \end{pmatrix} \qquad \hat{\theta}_2 = \begin{pmatrix} 1 & & & \\ & -1 & & \\ & & 1 & \\ & & & -1 \end{pmatrix} \qquad \text{(C.14)}$$

$$\hat{\psi} = \begin{pmatrix} 1 & & & \\ & -1 & & \\ & & -1 & \\ & & & 1 \end{pmatrix} \qquad \hat{\varphi} = \begin{pmatrix} 1 & & & \\ & 1 & & \\ & & 1 & \\ & & & 1 \end{pmatrix}$$

where we have two single-qubit phases $\hat{\theta}_1 = \sigma_z \otimes I$ and $\hat{\theta}_2 = I \otimes \sigma_z$, a two-qubit phase $\hat{\psi} = \sigma_z \otimes \sigma_z$ and a global phase $\hat{\varphi} = I \otimes I$.

Acquired phases During the σ_z geometric phase gate we acquire phases due to two different mechanisms: a geometric phase $\Phi = \text{Im}\left(\int_\gamma \alpha^* d\alpha\right)$ acquired due to the trajectory through motional phase space and a phase θ_{LS} caused by a time-dependent light-shift.

The geometric phase has a two-qubit and a single-qubit phase component $\Phi = \Psi + \Theta$. For a maximally entangling gate we require $\Psi = \pi$. For fast gates we need to sum over the phase acquired by the different motional modes, which for very fast gates have opposite signs. In general Ψ depends on ϕ_0 for fast gates, we therefore need to specifically design sequences such that Ψ is independent of ϕ_0.
The single-qubit geometric phase Θ is caused by asymmetry in the Rabi frequencies of different spin states and increases with lower gate efficiencies. This phase can be made to cancel by implementing a 2-loop gate and flipping all spins in the middle of the gate using a π-pulse. For fast gates this is undesirable, as it would increase the gate length. Instead the phase is cancelled by shifting the last $\pi/2$-pulse of the surrounding Ramsey interferometer by the corresponding phase $\phi_{\text{sq}} = -\Theta$. For mixed species the single-qubit phase on the different species has in general different magnitude. Suppressing the variance of Ψ due to random ϕ_0 also reduces the variance of Θ.

The light shift phase θ_{LS} is purely a single-qubit phase. It is caused by coupling of the Raman lasers to the 'carrier' of the qubit level. It could be cancelled by placing the gate sequence inside a spin-echo sequence with an appropriate phase-shift at the centre, however not at the same time as cancelling the geometric single-qubit phase. Instead, for slow gates, it is suppressed by smoothing the edges of the laser pulse to reduce higher frequency components. For fast gates this is not possible, since the time-constant of shaping required to sufficiently suppress errors would typically be longer than an entire fast gate sequence. Instead, the timings of the fast gate sequences are chosen such that $\theta_{LS} = 0$.

Control phases There are several external control phases set and/or influenced by the experimental apparatus that affect the gate operation. One is ϕ_z, the phase difference at which two ions experience the Raman beams. It can be set by adjusting the ions spacing, and is $e^{i\phi_z} = -1$ in this thesis. The phase offset between the Rabi frequency on $|\uparrow\rangle$ and $|\downarrow\rangle$ can be set by adjusting the polarisation of the Raman beams. It is also set to -1 in this thesis work. The phase between the two Raman beams ϕ_0 is not controlled and varies from shot to shot of the experiment. Although both Raman beams originate from the same laser, they traverse different paths on the optics table and vibration of mirrors and other optics cause the relative phase to fluctuate. These noise sources are all at acoustic frequencies or lower and therefore ϕ_0 is stable over the duration of a single gate.

Phase noise in the Raman lasers or phase chirps caused by the AOMs can also modulate the phase of the driving field and cause errors.

References

1. Magnus W (1954) On the exponential solution of differential equations for a linear operator. Commun Pure Appl Math VII:649–673
2. Blanes S, Casas F, Oteo JA, Ros J (2009) The Magnus expansion and some of its applications. Phys Reports 470:151–238. ISSN: 03701573
3. Wünsche A (1991) Displaced Fock states and their connection to quasiprobabilities. Quantum Opt 3:359–383
4. Wineland DJ et al (1998) Experimental issues in coherent quantum-state manipulation of trapped atomic ions. J Res Natl Inst Standards Technol 103:259–328

Appendix D
^{88}Sr$^+$ Properties

D.1 Einstein A Coefficients

Transition	A_{ik}	Source	Theory	Source
$A_{5S_{1/2}-5P_{3/2}}$	$142 \times 10^6\,\mathrm{s}^{-1}$	[1]	$141.29 \times 10^6\,\mathrm{s}^{-1}$	[2]
	$143(6) \times 10^6\,\mathrm{s}^{-1}$	[3]		
$A_{5S_{1/2}-5P_{1/2}}$	$127 \times 10^6\,\mathrm{s}^{-1}$	[1]	$128.04 \times 10^6\,\mathrm{s}^{-1}$	[2]
	$127(5) \times 10^6\,\mathrm{s}^{-1}$	[3]		
	$127.9(1.3) \times 10^6\,\mathrm{s}^{-1}$	[4]		
$A_{4D_{3/2}-5P_{3/2}}$	$1.0(2) \times 10^6\,\mathrm{s}^{-1}$	[3]	$0.96 \times 10^6\,\mathrm{s}^{-1}$	[2]
$A_{4D_{3/2}-5P_{1/2}}$	$9.5(2.0) \times 10^6\,\mathrm{s}^{-1}$	[3]	$7.54 \times 10^6\,\mathrm{s}^{-1}$	[2]
	$7.46(14) \times 10^6\,\mathrm{s}^{-1}$	[4]		
$A_{4D_{5/2}-5P_{3/2}}$	$8.7(1.5) \times 10^6\,\mathrm{s}^{-1}$	[3]	$8.06 \times 10^6\,\mathrm{s}^{-1}$	[2]

D.2 Energies Relative to $4S_{1/2}$

$k_{5P_{3/2}} = 24516.65 \frac{1}{\mathrm{cm}}$ [5]
$k_{5P_{1/2}} = 23715.19 \frac{1}{\mathrm{cm}}$ [5]
$k_{4D_{5/2}} = 14836.24 \frac{1}{\mathrm{cm}}$ [5]
$k_{4D_{3/2}} = 14555.90 \frac{1}{\mathrm{cm}}$ [5].

© Springer Nature Switzerland AG 2020
V. M. Schäfer, *Fast Gates and Mixed-Species Entanglement with Trapped Ions*,
Springer Theses, https://doi.org/10.1007/978-3-030-40285-3

D.3   Wavelengths

*all wavelengths quoted are in vacuum (converted from cited sources with $\lambda_{vac} = n_\lambda \lambda_{air}$).

$\lambda_{5S_{1/2}-5P_{3/2}}$	407.8865(1) nm	$\Rightarrow$	$\nu_{408} = 734, 989.88(18)$ GHz	[6, 5]
	407.89 nm			[7,8]
$\lambda_{5S_{1/2}-5P_{1/2}}$	421.6712(1) nm	$\Rightarrow$	$\nu_{422} = 710, 962.69(18)$ GHz	[6, 5]
	421.67 nm			[7,8]
$\lambda_{4D_{3/2}-5P_{1/2}}$	1091.8 nm			[9]
	1091.7864(3) nm	$\Rightarrow$	$\nu_{1092} = 274, 588.94(8)$ GHz	[6, 5]
$\lambda_{5S_{1/2}-4D_{5/2}}$	674.0 nm			[10]
	674.02559 nm	$\Leftarrow$	$\nu_{674} = 444, 779.044095$ GHz	[11]
$\lambda_{5S_{1/2}-4D_{3/2}}$	687.0 nm			[10]
	687.0066(5) nm	$\Rightarrow$	$\nu_{687} = 436, 374.94(32)$ GHz	[6, 5]
$\lambda_{5P_{3/2}-4D_{3/2}}$	1003.9 nm			[10]
	1003.9406(3) nm	$\Rightarrow$	$\nu_{1004} = 298, 615.73(8)$ GHz	[6, 5]
$\lambda_{5P_{3/2}-4D_{5/2}}$	1033.0 nm			[10, 8]
	1033.0140(3) nm	$\Rightarrow$	$\nu_{1033} = 290, 211.43(8)$ GHz	[6, 5]

D.4   Lifetimes

$\tau_{5P_{3/2}} = 6.63(7)$ ns [7]
$\tau_{5P_{1/2}} = 7.39(7)$ ns [7]
$\tau_{4D_{5/2}} = 390.8(1.6)$ ms [12]
$\tau_{4D_{3/2}} = 435(4)$ ms [10].

D.5   Branching Ratios

*inferred from measured lifetimes and measured Einstein A coefficients with $B_r = A_{ij} \cdot \tau$

Transition	Branching ratio (%)	Source
$5P_{3/2} - 4D_{3/2}$	0.66(13)	[7] + [3]
$5P_{3/2} - 4D_{5/2}$	5.8(1.0)	[7] + [3]
$5P_{3/2} - 4S_{1/2}$	94.8(4.1)	[7] + [3]
$5P_{1/2} - 4D_{3/2}$	5.51(12)	[7] + [4]
$5P_{1/2} - 4S_{1/2}$	94.5(1.3)	[7] + [4]

References

1. Sansonetti JE, Martin WC (2005) Strontium atomic spectroscopic data. J Phys Chem Ref Data 34:2018–2021
2. Jiang D, Arora B, Safronova MS, Clark CW (2009) Blackbody-radiation shift in a 88Sr+ ion optical frequency standard. J Phys B: Atomic Mol Opt Phys 42:154020. ISSN: 0953-4075
3. Gallagher A (1967) Oscillator strengths of Ca II, Sr II, and Ba II. Phys Rev 157, 24–30. ISSN: 0031899X
4. Likforman J-P, Tugayé V, Guibal S, Guidoni L (2016) Precision measurement of the branching fractions of the 5p 2P1/2 state in 88 Sr+ with a single ion in a microfabricated surface trap. Phys Rev A 93, 052507. ISSN: 2469-9926
5. Sansonetti JE (2012) Wavelengths, transition probabilities, and energy levels for the spectra of Strontium ions (Sr II through Sr XXXVIII). J Phys Chem Ref Data 41, 013102. ISSN: 00472689
6. Kramida A, Ralchenko Y, Reader J (2017) NIST Atomic Spectra Database (ver. 5.5.1) 2017. https://physics.nist.gov/asd
7. Pinnington EH, Berends RW, & Lumsden M (1995) Studies of laser-induced fluorescence in fast beams of Sr+ and Ba+ ions. J Phys B: Atomic Mol Opt Phys 28:2095–2103. ISSN: 0953-4075. 218
8. Madej AA, Bernard, JE, Dubé P, Marmet L, Windeler RS (2004) Absolute frequency of the 88 Sr+ 5s 2S1/2-4d 2D5/2 reference transition at 445 THz and evaluation of systematic shifts. Phys Rev A 70, 012507. ISSN: 10502947
9. Madej AA, Berger WE, Hanes GR, O'Sullivan MS (1989) Tunable frequency narrowed Nd3+-doped fiber laser for excitation of the 5p 2P1/2-4d 2D3/2 transition in Sr+. Opt Commun 73:147–152
10. Mannervik S et al (1999) Lifetime measurement of the metastable 4d 2D3/2 level in Sr+ by optical pumping of a stored ion beam. Phys Rev Lett 83:698–701. ISSN: 0031-9007
11. Margolis HS et al (2003) Absolute frequency measurement of the 674-nm 88Sr+ clock transition using a femtosecond optical frequency comb. Phys Rev A 67, 032501. ISSN: 1050-2947
12. Letchumanan V, Wilson MA, Gill P, Sinclair AG (2005) Lifetime measurement of the metastable 4d 2D5/2 state in 88Sr+ using a single trapped ion. Phys Rev A 72, 012509. ISSN: 1050-2947

Printed in the United States
By Bookmasters